Numerical Calculus

Numerical Calculus

Approximations, Interpolation, Finite Differences, Numerical Integration, and Curve Fitting

By William Edmund Milne

1949

PRINCETON UNIVERSITY PRESS

PRINCETON, NEW JERSEY

LONDON: GEOFFREY CUMBERLEGE, OXFORD UNIVERSITY PRESS

SECOND PRINTING, 1950

PRINTED IN THE UNITED STATES OF AMERICA

PREFACE

The growth of computational facilities in the period since World War I has already been phenomenal, and the possibilities in the near future are beyond imagining. The number of excellent calculating machines now available in almost every office or laboratory, to say nothing of the amazing Sequence Controlled Calculator, or the equally marvelous Electronic Numerical Integrator and Calculator, makes it possible to solve whole categories of problems that only yesterday were prohibitively difficult. In consequence the subject of numerical analysis is surely destined to make enormous strides in the decades to come.

On the other hand our traditional courses in college and graduate mathematics too often turn out students poorly trained in the art of translating theoretical analysis into the concrete numerical results generally required in practical applications. As a physicist friend of mine once said, "You mathematicians know *how* to solve this problem but you can't actually *do* it".

The aim of this book is to aid in bridging the considerable gulf between classroom mathematics and the numerical applications. It is designed to provide rudimentary instruction in such topics as solution of equations, interpolation, numerical integration, numerical solution of differential equations, finite differences, approximations by Least Squares, smoothing of data, and simple equations in finite differences. The presentation is intentionally elementary so that anyone with some knowledge of calculus and differential equations can read it understandingly. Mathematical elegance and rigor have frequently been sacrificed in favor of a purely naive treatment.

PREFACE

From the wealth of material available the decision to include or to exclude specific topics has posed a vexing problem, and personal interest rather than objective logic has doubtless influenced the choice. It is hoped that enough has been included to meet a fairly general need and to stimulate wider study of this interesting and useful branch of mathematics.

The author is indebted to Mr. James Price for the calculation of Table VI, to Professor William M. Stone for some of the material in Article 82, and to Professor Burns W. Brewer for criticisms of Chapter VIII.

William Edmund Milne, Ph.D., D.Sc.
Professor of Mathematics
Oregon State College

Corvallis, Oregon
March 1948

CONTENTS

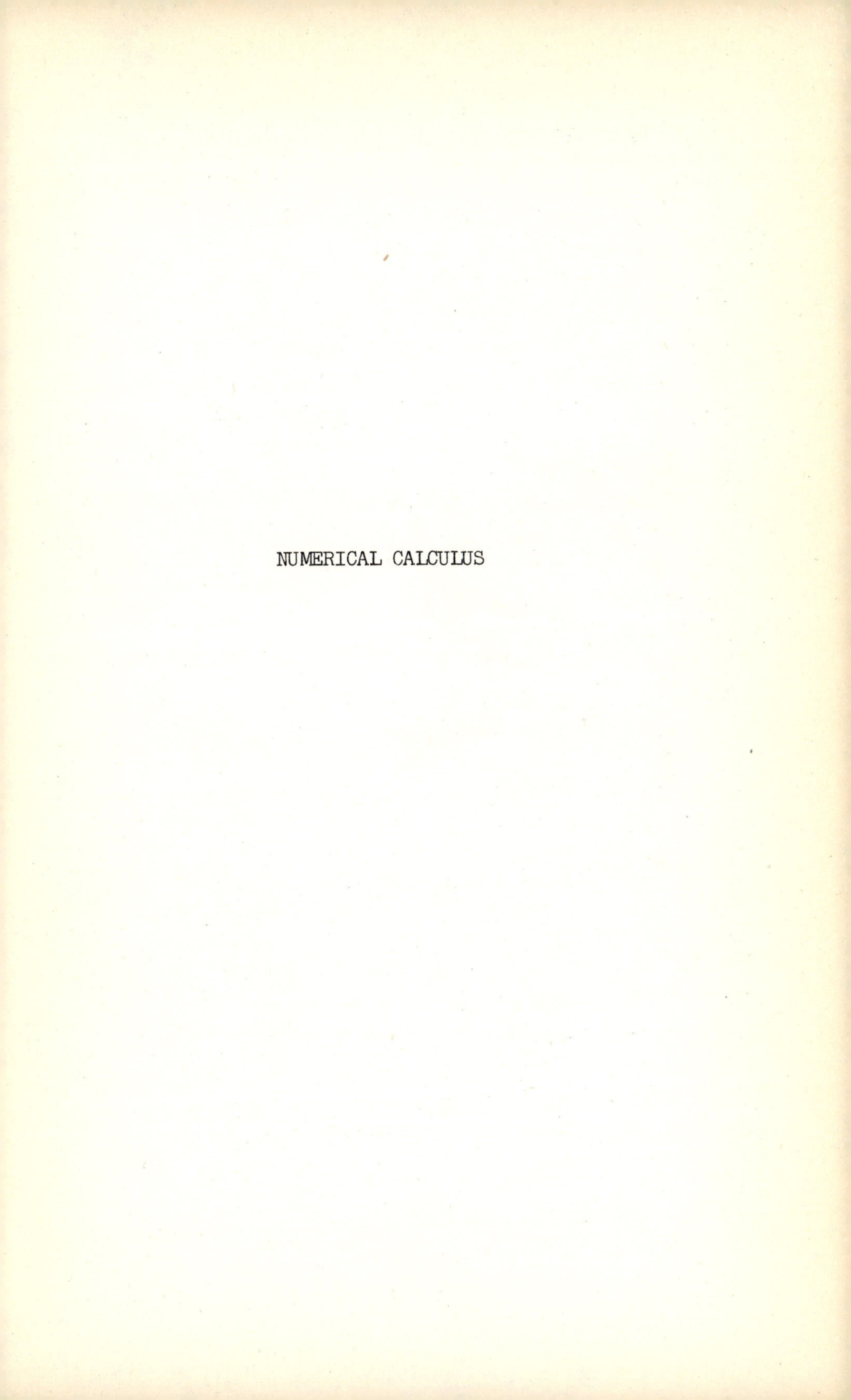

NUMERICAL CALCULUS

Chapter I
SIMULTANEOUS LINEAR EQUATIONS

A great variety of problems in pure mathematics and in the several branches of applied mathematics involve either directly or indirectly sets of simultaneous linear equations. Hence, it is appropriate to devote this first chapter to methods of solving such equations and to an estimate of the attainable accuracy of the solution.

For theoretical investigations and for the case of equations with literal coefficients we commonly employ determinants. For the case of numerical coefficients we use a systematic scheme of elimination adapted to machine calculations. While the two methods are not fundamentally distinct, they differ so much in details of operation that it is convenient to treat them separately. However, since a knowledge of determinants is almost essential for a full understanding of the method of elimination it is not practical to make the two treatments entirely independent.

1. DETERMINANTS

We present here without proof a brief account of the elementary properties of determinants.

(a) The value of a two-rowed determinant is given by the formula,

$$\begin{vmatrix} a_1 & b_1 \\ a_2 & b_2 \end{vmatrix} = a_1b_2 - a_2b_1$$

(b) The value of a three-rowed determinant is given by the formula,

$$\begin{vmatrix} a_1 & b_1 & c_1 \\ a_2 & b_2 & c_2 \\ a_3 & b_3 & c_3 \end{vmatrix} = a_1b_2c_3 + a_2b_3c_1 + a_3b_1c_2 - a_3b_2c_1 - a_2b_1c_3 - a_1b_3c_2.$$

(c) The determinant of n-1 rows obtained from an n-rowed determinant by striking out the i-th row and j-th column is called the <u>minor</u> of the element in the i-th row and the j-th column. Thus in the determinant

$$(1) \qquad D = \begin{vmatrix} a_1 & b_1 & c_1 & d_1 \\ a_2 & b_2 & c_2 & d_2 \\ a_3 & b_3 & c_3 & d_3 \\ a_4 & b_4 & c_4 & d_4 \end{vmatrix}$$

the minor of c_2 is

$$\begin{vmatrix} a_1 & b_1 & d_1 \\ a_3 & b_3 & d_3 \\ a_4 & b_4 & d_4 \end{vmatrix}$$

In a similar manner by striking out two rows and two columns we obtain minors of order two less than the original determinant, by striking out three rows and three columns we obtain minors of order three less, etc.

(d) By the <u>cofactor</u> of an element in the i-th row and j-th column is meant the minor of that element multiplied by $(-1)^{i+j}$. It will be convenient to designate elements by small letters and their corresponding cofactors by the same letter capitalized. Thus from (1) we have

$$C_2 = - \begin{vmatrix} a_1 & b_1 & d_1 \\ a_3 & b_3 & d_3 \\ a_4 & b_4 & d_4 \end{vmatrix}, \quad B_4 = + \begin{vmatrix} a_1 & c_1 & d_1 \\ a_2 & c_2 & d_2 \\ a_3 & c_3 & d_3 \end{vmatrix}, \text{ etc.}$$

(e) The sum of products of each element of a row or

(or column) by its own cofactor is equal to the determinant. For example, from (1)

$$D = a_2A_2 + b_2B_2 + c_2C_2 + d_2D_2.$$

(f) The sum of products of each element of a row (or column) by the corresponding cofactors of another row (or column) is equal to zero. Thus

$$a_2A_3 + b_2B_3 + c_2C_3 + d_2D_3 = 0.$$

(g) If two columns (or two rows) of a determinant are interchanged, the value of the determinant is changed in sign.

(h) If corresponding elements of two columns (or two rows) of a determinant are equal, the value of the determinant is zero.

(i) If corresponding elements of two columns (or two rows) of a determinant are proportional, the value of the determinant is zero.

(j) The value of a determinant is unchanged if

(1) Rows and columns are interchanged without changing the order in which they occur.

(2) A factor is removed from all elements of a row (or column) and the resulting determinant is multiplied by this factor.

(3) The elements of a row (or column) are changed by the addition of a constant times the corresponding element of another row (or column). Thus,

$$\begin{vmatrix} a_1 & b_1 & c_1 \\ a_2 & b_2 & c_2 \\ a_3 & b_3 & c_3 \end{vmatrix} = \begin{vmatrix} (a_1 + kc_1) & b_1 & c_1 \\ (a_2 + kc_2) & b_2 & c_2 \\ (a_3 + kc_3) & b_3 & c_3 \end{vmatrix}$$

(k) The number of rows (or columns) is called the order of the determinant.

(1) Rank of a determinant. If the value of a determinant of n-th order is not zero the determinant is said to be of rank n. If the determinant is zero the order of the non-vanishing minor of highest order is said to be the rank of the determinant.

Exercises

Evaluate by (b)

1. $\begin{vmatrix} 2 & 3 & 7 \\ -4 & 2 & 1 \\ 3 & 6 & 5 \end{vmatrix}$ 2. $\begin{vmatrix} 4 & -1 & 2 \\ 2 & 3 & 2 \\ 6 & 2 & 4 \end{vmatrix}$ 3. $\begin{vmatrix} 3 & 6 & 5 \\ 2 & 1 & 2 \\ -1 & 1 & -1 \end{vmatrix}$

4. Evaluate 1, 2, 3 by cofactors of first row as in (e).
5. Evaluate 1, 2, 3 by cofactors of second column as in(e).
6. Verify (f) in 1, 2, 3, taking elements of first row by cofactors of second row.
7. Evaluate by (e)

$$\begin{vmatrix} 1 & 2 & 1 & 2 \\ 3 & 1 & 2 & 2 \\ 1 & 3 & 3 & 1 \\ 3 & 1 & 1 & 3 \end{vmatrix}$$

8. Evaluate 7 by method (j, 3).
9. Show that

$$\begin{vmatrix} 1 & x & x^2 \\ 1 & y & y^2 \\ 1 & z & z^2 \end{vmatrix} = (x - y)\ (y - z)\ (z - x).$$

10. Show that

$$\begin{vmatrix} 1 & x & x^2 & x^3 \\ 1 & a & a^2 & a^3 \\ 1 & b & b^2 & b^3 \\ 1 & c & c^2 & c^3 \end{vmatrix} = 0$$

is an algebraic equation of third degree in x whose roots are a, b, c.

11. Find the rank of the determinants in Exercises 1, 2, 3, 7.

2. SOLUTION OF LINEAR EQUATIONS BY DETERMINANTS

Let us consider a set of linear equations in n unknowns.

$$
(1) \qquad \begin{array}{l}
a_{11}x_1 + a_{12}x_2 + \dots + a_{1n}x_n = b_1 \\
a_{21}x_1 + a_{22}x_2 + \dots + a_{2n}x_n = b_2 \\
\dots\dots\dots\dots\dots\dots\dots\dots \\
a_{n1}x_1 + a_{n2}x_2 + \dots + a_{nn}x_n = b_n.
\end{array}
$$

Let the value of the determinant formed by the coefficients of the x's be denoted by D, and let the cofactor of the element a_{ij} in this determinant be A_{ij}. Now let us multiply the first equation above by A_{11}, the second by A_{21}, etc., and add the equations. The result of the addition, in view of the properties of the cofactors, is

$$
(2) \qquad Dx_1 = b_1A_{11} + b_2A_{21} + \dots + b_nA_{n1}.
$$

It is readily seen that the expression on the right is the expansion in elements of the first column of the determinant

$$
\begin{vmatrix}
b_1 & a_{12} & \dots & a_{1n} \\
b_2 & a_{22} & \dots & a_{2n} \\
\dots & \dots & \dots & \dots \\
b_n & a_{n2} & \dots & a_{nn},
\end{vmatrix}
$$

that is, it is the determinant D with the elements of the first column replaced by the b's. Let us denote this determinant by D_1, and in general let D_i denote the determinant obtained from D by replacing the i-th column by the column of b's. Then we have the equations,

$$
Dx_1 = D_1, \quad Dx_2 = D_2, \quad \dots, \quad Dx_n = D_n.
$$

Three cases are now to be distinguished.

I. $D \neq 0$. In this case the equations have <u>one and only one solution</u>, which is given by

$$(3) \qquad x_1 = \frac{D_1}{D}, \quad x_2 = \frac{D_2}{D}, \ . \ . \ ., \quad x_n = \frac{D_n}{D}$$

That this is actually a solution may be proved by direct substitution back into the original equations.

II. $D = 0$, not all $D_i = 0$. Then we have the impossible equation, $0 = D_i$, and the equations have <u>no solution</u>.

III. $D = 0$, all $D_i = 0$. Let r be the rank of D and r_i the rank of D_i. Then

(a) If any r_i is greater than r, there is no solution.

(b) If no r_i is greater than r, then r of the unknowns can be obtained in terms of the remaining n - r unknowns, as follows: Rearrange (if necessary) the order of the equations and the order of the unknowns in the equations so that the r-rowed determinant in the upper left hand corner of D (as rearranged) will not be zero. In the first r equations transpose to the right all the variables beyond the first r, and solve the first r equations for the first r unknowns as in Case I. The solutions thus found will satisfy the remaining equations also and will contain n - r undetermined variables.

<u>Matrix</u>. A system of mn quantities arranged in a rectangular array of m rows and n columns is called a <u>matrix</u>. In contrast to a determinant, which is a <u>quantity</u> determined from its matrix by certain rules, a matrix is not a quantity at all but an <u>array of quantities</u>. For example two determinants with different elements may have the same value, but two matrices with different elements are different matrices.

We shall not be concerned with the general properties of matrices, but we do find it convenient to define and use the term <u>rank</u> of a matrix. From any given matrix it is possible to set up determinants by striking out (if necessary) certain rows and columns from the matrix. The determinant of highest order that can be so formed has an order equal to the lesser of the two numbers m and n, if m and n are unequal, or equal to m if m and n are equal. The highest rank of the determinants of highest order is said to be the <u>rank</u> of the matrix.

The square matrix consisting of the array of coefficients in the left-hand members of equations (1) is called the <u>matrix of the coefficients</u>. The matrix

$$\begin{Vmatrix} a_{11} & a_{12} & \cdots & a_{1n} & b_1 \\ a_{21} & a_{22} & \cdots & a_{2n} & b_2 \\ \cdots & \cdots & \cdots & \cdots & \cdots \\ a_{n1} & a_{n2} & \cdots & a_{nn} & b_n \end{Vmatrix}$$

with n rows and n + 1 columns is called the <u>augmented matrix</u>.

Using this terminology we can express the results I, II, III above in the following way.

I. If the matrix of the coefficients is of rank n the equations have a unique solution.

II. If the rank of the augmented matrix is greater than the rank of the matrix of the coefficients the equations have no solutions.

III. If the ranks of the augmented matrix and of the matrix of the coefficients are both equal to n - k there are an infinite number of solutions expressible in terms of k arbitrary parameters.

<u>Exercises</u>

Determine whether the following equations have solutions, and obtain solutions when possible. Use determinants.

1. $$\begin{aligned} 2x - y + 3z &= 2. \\ x + y - z &= 3. \\ x - y + z &= 2. \end{aligned}$$

2. $$\begin{aligned} x + y + z &= 1. \\ x + 2y + 3z &= 0. \\ x + 4y + 9z &= 0. \end{aligned}$$

3. $$\begin{aligned} x + y + z &= 1. \\ x + 2y + 4z &= 0. \\ x + 3y + 9z &= 0. \end{aligned}$$

4. $$\begin{aligned} a - 2b + 4c - 8d + 16e &= 1. \\ a - b + c - d + e &= 2. \\ a &= 3. \\ a + b + c + d + e &= 3. \\ a + 2b + 4c + 8d + 16e &= 3. \end{aligned}$$

5. $$\begin{aligned} x + y + z &= 1. \\ 2x + 2y + z &= 0. \\ 3x - y + 2z &= 2. \end{aligned}$$

6. $$\begin{aligned} x + y + z &= 3. \\ 2x - 2y + z &= 1. \\ 3x - y + 2z &= 4. \end{aligned}$$

7. Solve the set of equations

$$\begin{aligned} x_1 + 1/2x_2 + 1/3x_3 + 1/4x_4 &= 1, \\ 1/2x_1 + 1/3x_2 + 1/4x_3 + 1/5x_4 &= 0, \\ 1/3x_1 + 1/4x_2 + 1/5x_3 + 1/6x_4 &= 0, \\ 1/4x_1 + 1/5x_2 + 1/6x_3 + 1/7x_4 &= 0. \end{aligned}$$

8. Solve the set of equations

$$\begin{aligned} 2x_1 - x_2 &= 1, \\ x_1 + 2x_2 - x_3 &= 1/2, \\ - x_2 + 2x_3 - x_4 &= 1/3, \\ - x_3 + 2x_4 &= 1/4. \end{aligned}$$

9. Solve the set of equations

$$\begin{aligned} x_o + x_1 + x_2 + x_3 + x_4 &= 1, \\ x_o + x_1 + 2x_2 + 3x_3 + 4x_4 &= 0, \\ x_o + x_1 + 2x_2 + 6x_3 + 12x_4 &= 0, \\ x_o + x_1 + 2x_2 + 6x_3 + 24x_4 &= 0, \\ x_o - x_1 - x_2 - x_3 - x_4 &= 0. \end{aligned}$$

10. Show that equations of the type

$$
\begin{aligned}
&x_1 &&= c_1\\
&a_{12}x_1 + x_2 &&= c_2\\
&a_{13}x_1 + a_{23}x_2 + x_3 &&= c_3\\
&a_{14}x_1 + a_{24}x_2 + a_{34}x_3 + x_4 &&= c_4\\
&\cdots\cdots\cdots\cdots
\end{aligned}
$$

always have a unique solution, which can be found by simple substitutions.

11. Show that the solution of the equations in Exercise 10 may be written

$$
\begin{aligned}
x_1 &= c_1,\\
x_2 &= c_2 - a_{12}c_1,\\
x_3 &= c_3 - a_{23}c_2 + \begin{vmatrix} a_{12} & 1 \\ a_{13} & a_{23} \end{vmatrix} c_1,\\
x_4 &= c_4 - a_{34}c_3 + \begin{vmatrix} a_{23} & 1 \\ a_{24} & a_{34} \end{vmatrix} c_2 - \begin{vmatrix} a_{12} & 1 & 0 \\ a_{13} & a_{23} & 1 \\ a_{14} & a_{24} & a_{34} \end{vmatrix} c_1,\\
&\cdots\cdots\cdots\cdots
\end{aligned}
$$

3. HOMOGENEOUS EQUATIONS

A set of n simultaneous linear equations with n unknowns of the form

$$
\begin{aligned}
a_{11}x_1 + a_{12}x_2 + \cdots + a_{1n}x_n &= 0,\\
a_{21}x_1 + a_{22}x_2 + \cdots + a_{2n}x_n &= 0,\\
\cdots\cdots\cdots\cdots\\
a_{n1}x_1 + a_{n2}x_2 + \cdots + a_{nn}x_n &= 0,
\end{aligned}
$$

in which the constant terms are all zero always has at

least one solution, namely $x_1 = x_2 = \dots = x_n = 0$. If the determinant

$$D = \begin{vmatrix} a_{11} & \dots & a_{1n} \\ \dots & \dots & \dots \\ a_{n1} & \dots & a_{nn} \end{vmatrix}$$

is not zero the equations have no other solution. If D is of rank r where $r < n$, it is possible to solve for r of the variables in terms of the remaining $n - r$ variables.

Example 1. Investigate the equations

$$\begin{aligned} 2x + y - z &= 0, \\ x + 2y + z &= 0, \\ -x + y + 2z &= 0. \end{aligned}$$

Here we find that the determinant D is of rank 2. We may therefore transpose one variable, say z, to the right side, and solve for x and y, obtaining

$$x = z, \quad y = -z$$

If the rank of D is $n - 1$, not all of the cofactors of D are zero. In this case a more symmetrical way to express the non-zero solutions of the homogeneous equations is to set each x_j proportional to the cofactor of the j-th element of one row of D (provided the cofactors of this row are not all zero). Thus

$$x_1 = kA_{i1}, \quad x_2 = kA_{i2}, \quad \dots, \quad x_n = kA_{in},$$

where k is an arbitrary constant. That these values actually furnish solutions follows at once from (e) and (f) of Article 1, since in this case $D = 0$.

Applying this procedure to the equations of Example 1, we get

$$x = 3k, \quad y = -3k, \quad z = 3k,$$

or if we let $3k = k'$

$$x = k', \quad y = -k', \quad z = k'.$$

This result is seen to be equivalent to the one previously obtained.

A type of problem that arises frequently in a wide variety of applications is illustrated by the following example:

Example 2. For what value of the parameter λ will the equations

$$\begin{aligned} 4x + 2y + z &= \lambda x \\ 2x + 4y + 2z &= \lambda y \\ x + 2y + 4z &= \lambda z \end{aligned}$$

have non-zero solutions.

Collecting the coefficients of x, y, and z, we obtain as the determinant D of the homogeneous equations

$$D = \begin{vmatrix} (4 - \lambda) & 2 & 1 \\ 2 & (4 - \lambda) & 2 \\ 1 & 2 & (4 - \lambda) \end{vmatrix}$$

In order that the homogeneous equations have a solution not zero the determinant D must vanish. Upon expanding the determinant we obtain the following cubic equation which the parameter λ must satisfy:

$$(4 - \lambda)^3 - 9(4 - \lambda) + 8 = 0.$$

Solving this equation for λ, we find three values $\lambda_1 = 1.62772$, $\lambda_2 = 3$, $\lambda_3 = 7.37228$, for each of which the given equations have a set of non-zero solutions.

Exercises

1. Solve

$$\begin{aligned} 3x + 2y - z &= 0, \\ 2x - y + 2z &= 0, \\ x - 4y + 5z &= 0. \end{aligned}$$

2. Solve

$$\begin{aligned} 2x_1 + x_2 - x_3 + x_4 &= 0, \\ x_1 - x_2 - x_3 + x_4 &= 0, \\ x_1 - 4x_2 - 2x_3 + 2x_4 &= 0, \\ 5x_1 + x_2 - 3x_3 + 3x_4 &= 0. \end{aligned}$$

3. Determine λ and find the corresponding non-zero solutions:

$$\begin{aligned} 3x - y &= \lambda x, \\ -x + 3y &= \lambda y. \end{aligned}$$

4. Determine λ and find the corresponding non-zero solutions:

$$\begin{aligned} \lambda x &= 7x + 3y, \\ \lambda y &= 3x + 7y + 4z, \\ \lambda z &= \phantom{3x + {}} 4y + 7z. \end{aligned}$$

5. Show that the equations

$$\begin{aligned} \lambda x &= x \cos \theta - y \sin \theta \\ \lambda y &= x \sin \theta + y \cos \theta \qquad \sin \theta \neq 0. \end{aligned}$$

have no real non-zero solutions.

4. THE METHOD OF ELIMINATION

The use of determinants for the solution of linear equations becomes excessively cumbrous if the number of equations exceeds four or five and especially so if the coefficients are numbers expressed in several digits. Hence for practical calculation we resort to a systematic method of elimination. The explanation is given for the case of four equations with four unknowns, as the reader can readily extend the procedure to the general case. It should be kept in mind that the detailed steps in the following discussion are given to make clear the basis for the short process shown in the next Article, and are not intended as a pattern for actual computation.

Let the set of equations to be solved be written as follows:

$$
\begin{aligned}
&(1) \qquad a_{11}x_1 + a_{12}x_2 + a_{13}x_3 + a_{14}x_4 = a_{15},\\
&(2) \qquad a_{21}x_1 + a_{22}x_2 + a_{23}x_3 + a_{24}x_4 = a_{25},\\
&(3) \qquad a_{31}x_1 + a_{32}x_2 + a_{33}x_3 + a_{34}x_4 = a_{35},\\
&(4) \qquad a_{41}x_1 + a_{42}x_2 + a_{43}x_3 + a_{44}x_4 = a_{45},
\end{aligned}
$$

where a_{11} is not zero. (If the coefficient of x_1 in the first equation is zero, we rearrange the order of the equations choosing for the first equation one in which the coefficient of x_1 is not zero.)

The first stage of the elimination is effected by the following steps:

a) Divide equation (1) by a_{11};

b) Multiply the resulting equation first by a_{21} and subtract from equation (2); then by a_{31} and subtract from (3); then by a_{41} and subtract from (4). The result is the set of equations

$$x_1 + b_{12}x_2 + b_{13}x_3 + b_{14}x_4 = b_{15}, \tag{5}$$

$$b_{22}x_2 + b_{23}x_3 + b_{24}x_4 = b_{25}, \tag{6}$$

$$b_{32}x_2 + b_{33}x_3 + b_{34}x_4 = b_{35}, \tag{7}$$

$$b_{42}x_2 + b_{43}x_3 + b_{44}x_4 = b_{45}, \tag{8}$$

where the b's are obtained from the a's by the formulas

$$b_{1j} = a_{1j}/a_{11}, \quad (j = 2, 3, 4, 5) \tag{9}$$

$$b_{ij} = a_{ij} - a_{i1}b_{1j}, \quad (i = 2, 3, 4), \quad (j = 2, 3, 4, 5). \tag{10}$$

To perform the second stage of the elimination we operate on equations (6), (7), and (8) in precisely the same manner as we did in the case of equations (1), (2), (3), and (4), obtaining

$$x_2 + c_{23}x_3 + c_{24}x_4 = c_{25}, \tag{11}$$

$$c_{33}x_3 + c_{34}x_4 = c_{35}, \tag{12}$$

$$c_{43}x_3 + c_{44}x_4 = c_{45}, \tag{13}$$

where

$$c_{2j} = b_{2j}/b_{22}, \quad (j = 3, 4, 5), \tag{14}$$

$$c_{ij} = b_{ij} - b_{i2}c_{2j}, \quad (i = 3, 4), \quad (j = 3, 4, 5). \tag{15}$$

The third stage is performed on equations (12) and (13) giving

$$x_3 + d_{34}x_4 = d_{35}, \tag{16}$$

$$d_{44}x_4 = d_{45}, \tag{17}$$

where

$$d_{3j} = c_{3j}/c_{33}, \quad (j = 4, 5). \tag{18}$$

$$d_{ij} = c_{ij} - c_{i3}d_{3j}, \quad (i = 4), \quad (j = 4, 5). \tag{19}$$

The fourth stage consists simply in dividing (17) by d_{44}, giving

(20) $$x_4 = e_{45},$$

in which $e_{45} = d_{45}/d_{44}$.

The values of the x's are now found in turn from equations (20), (16), (11), and (5), as follows:

(21) $$x_4 = e_{45},$$

(22) $$x_3 = d_{35} - d_{34}x_4,$$

(23) $$x_2 = c_{25} - c_{24}x_4 - c_{23}x_3,$$

(24) $$x_1 = b_{15} - b_{14}x_4 - b_{13}x_3 - b_{12}x_2.$$

Simple substitutions complete the solution by elimination.

In case it is necessary to solve a set of equations without the aid of a calculating machine, it may be best to follow the steps outlined above.

5. NUMERICAL SOLUTION OF LINEAR EQUATIONS

If a calculating machine is used it is possible to telescope the operations described in Article 4 to such an extent that the entire computation, including the original equations and the final values of the x's, can be compressed into the following compact form:

(A)
$$\begin{matrix} a_{11} & a_{12} & a_{13} & a_{14} & a_{15} \\ a_{21} & a_{22} & a_{23} & a_{24} & a_{25} \\ a_{31} & a_{32} & a_{33} & a_{34} & a_{35} \\ a_{41} & a_{42} & a_{43} & a_{44} & a_{45} \end{matrix}$$

	a_{11} [1]	b_{12} [2]	b_{13} [2]	b_{14} [2]	b_{15} [2]
	a_{21} [1]	b_{22} [3]	c_{23} [4]	c_{24} [4]	c_{25} [4]
(B)	a_{31} [1]	b_{32} [3]	c_{33} [5]	d_{34} [6]	d_{35} [6]
	a_{41} [1]	b_{42} [3]	c_{43} [5]	d_{44} [7]	e_{45} [8]

(C)	x_1 [12]	x_2 [11]	x_3 [10]	x_4 [9]

The matrix (A) is the augmented matrix of equations (1) to (4) Article 4. The matrix (B) is seen to consist of numbers occurring at various stages of the elimination. The matrix (C), consisting of one row only, gives the values of the x's. The numbers in brackets in (B) and (C) are inserted solely to indicate the order of operations and to facilitate the explanation.

[1] Copy the first column of (A).

[2] Divide the first row of (A) (excepting a_{11}) by a_{11}.

[3] Compute by the formula

$$b_{j2} = a_{j2} - a_{j1}b_{12}.$$

Note that the a_{j2} comes from (A) while the a_{j1} comes from the row and b_{12} from the column of (B) in which the desired b_{j2} lies. The a_{j2} is put in the machine and then the multiplication of a_{j1} by b_{12} is done with either the + bar or - bar according to the actual sign, leaving b_{j2} (or its complement) on the dials of the machine. The number is then recorded in the proper place in (B).

[4] Compute by the formula

$$c_{2j} = (a_{2j} - a_{21}b_{1j})/b_{22}$$

The numerator is found by a process similar to [3]. While the number is still on the dials the division by b_{22} is performed and the result recorded.

[5] Compute by the formula

$$c_{13} = a_{13} - a_{11}b_{13} - b_{12}c_{23}$$

This is done on the machine just as in [3] except that there are now two multiplications.

[6] Compute by the formula

$$d_{3j} = (a_{3j} - a_{31}b_{1j} - b_{32}c_{2j})/c_{33}.$$

By this time the careful reader will detect the rule for the calculation of the elements of (B). It may be stated as follows:

Rule. To find any element in (B)

a) Take the corresponding element in (A).

b) From it subtract the products of the elements of (B) in the row to the left and in the column above the desired element, taking the products in order, i.e. first in row by first in column, second in row by second in column, etc. (Of course the actual signs of these products must be watched. We may have arithmetical additions instead of subtractions.)

c) If the desired element is on or below the principal diagonal (i.e. the diagonal containing a_{11}, b_{22}, c_{33}, etc.) record the result obtained in b). If the desired element is to the right of the diagonal divide the result obtained in b) by the diagonal element in the same row.

We have now completed the calculation of (B). The entries in (C) are obtained in order from formulas (21), (22), (23), and (24) of Article 4. Referring to these formulas, we note that all the coefficients involved in them have already been computed in (B). We also note that each x is calculated by a continuous operation on the machine just as were the elements of (B). They are calculated in the order indicated by the numerals [9], [10], [11], [12].

The essential simplicity of the foregoing process becomes more evident when it is applied to a numerical example. The complete solution of the following set of equations is shown below:

	x_1		x_2		x_3		x_4			
	6.4375	+	2.1849	-	3.7474	+	1.8822	=	4.6351	(4.635100)
	2.1356	+	5.2101	+	1.5220	-	1.1234	=	5.2131	(5.213101)
(A)	-3.7362	+	1.4998	+	7.6421	+	1.2324	=	5.8665	(5.866499)
	1.8666	-	1.1104	+	1.2460	+	8.3312	=	4.1322	(4.132198)

	<u>6.4375</u>	0.339402	-0.582120	0.292381	0.720016
	2.1356	<u>4.485273</u>	0.616501	-0.389677	0.819445
(B)	-3.7362	2.767874	<u>3.760786</u>	0.904963	1.672125
	1.8666	-1.743928	3.407719	<u>4.022013</u>	-0.368189
(C)	2.185177	-0.560313	2.005322	-0.368189	

<u>Remarks</u>: 1. When the x's are obtained they are checked by substitution into the left-hand members of (A). The values thus found are recorded in parentheses in the 6th column at the right of (A) and are seen to agree closely with the actual right-hand members in column (5).

2. In performing the sequence of multiplications required in computing the elements of (B) the following device has been found useful. Instead of recording the entries below the diagonal in (B) as shown above, let us record them in a column on a separate slip of paper, arranged so that when this slip is properly placed beside a given column of (A) and (B) all operations are clearly indicated and pairs of numbers to be multiplied stand side by side. The slip for computing the fourth row of (B) might be set up as follows:

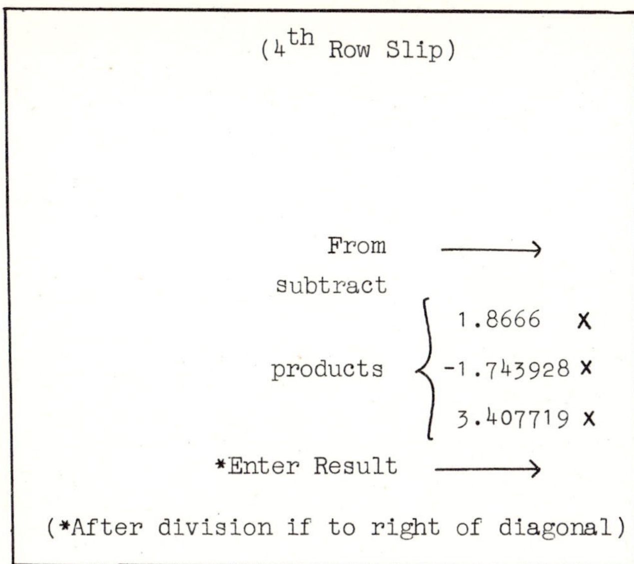

In using such a slip to compute the entry in the fourth row and fourth column of (B) for example we lay the slip on the computation sheet just to the left of the fourth column. The slip and column then appear as shown below

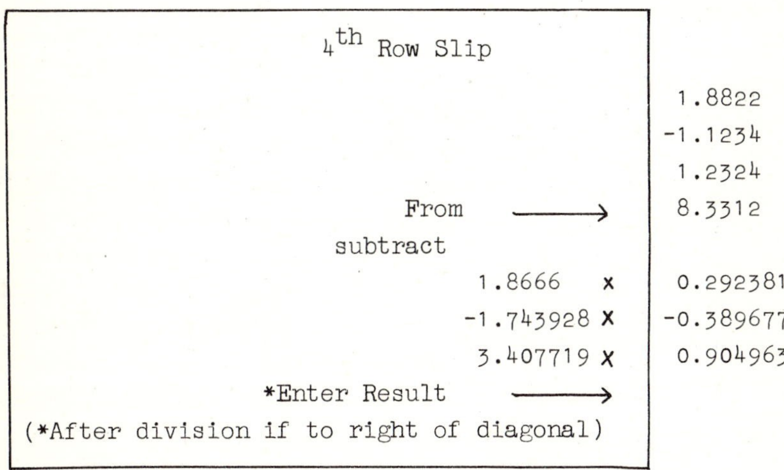

With this device we practically eliminate any possibility of taking the wrong entry out of (A) or of using a wrong

multiplier by losing our place in the matrix (B). For the slip shows us that we are to perform on the machine the operations

$$8.3312-(1.8666)(0.292381)-(-1.743929)(-0.389677)$$
$$-(3.407719)(0.904963),$$

which yield the result 4.022013.

Note that the first entry on the slip for the fourth row is the number in the first column and fourth row of (A). To get the second entry on the slip we place the slip against the second column of (A) and (B) and find that we have to compute

$$-1.1104 - (1.8666)(0.339402) = -1.743928.$$

This is entered on the slip, then the slip is placed against the third column and the third entry is computed and recorded on the slip. The slip for the fourth row is now complete, and the remaining entries for the fourth row are recorded on (B) after calculation by means of the slip in the usual manner.

If the number of equations is not too great a single strip of paper can be used for all of the different rows, instead of having a separate piece for each row.

3. It is also convenient to use a separate strip of paper on which to record final values of the x's. Such a strip, when completed, would appear as shown below

X	X	X	X	
2.185177	-0.560313	2.005322	-0.368189	↑
x_1	x_2	x_3	x_4	

The entries are computed as follows: The value of x_4 is the last entry in column 5 of (B). This is recorded on the strip. Then the strip is laid just below the third row of

(B), so that each x is in its proper column. Then take the number to which the arrow points and from it subtract the product of x_4 by the number in (B) just above x_4. This gives

$$1.672125 - (-0.368189)(0.904963) = 2.005322 = x_3$$

This value of x_3 is recorded on the strip, which is then moved up to the second row of (B) for the calculation of x_2. The process is repeated until all the x's are found. The completed strip is also handy for the check substitutions into equations (A).

It may be remarked that the foregoing suggestions are given primarily to aid beginners. The experienced computer will devise methods of his own adapted to the particular problem at hand.

Exercises

Solve the following sets of equations.

1.
$$\begin{aligned} 4.215x_1 - 1.212x_2 + 1.105x_3 &= 3.216, \\ -2.120x_1 + 3.505x_2 - 1.632x_3 &= 1.247, \\ 1.122x_1 - 1.313x_2 + 3.986x_3 &= 2.112. \end{aligned}$$

2.
$$\begin{aligned} 36.47x_1 + 5.28x_2 + 6.34x_3 &= 12.26, \\ 7.33x_1 + 28.74x_2 + 5.86x_3 &= 15.15 \\ 4.63x_1 + 6.31x_2 + 26.17x_3 &= 25.22 \end{aligned}$$

3.
$$\begin{aligned} + 36x_2 + 71x_3 &= 100 \\ -36x_1 + 68x_3 &= 50 \\ -75x_1 - 70x_2 &= 0 \end{aligned}$$

4. $3421x_1 + 1234x_2 + 736x_3 + 124x_4 = 365$

$1202x_1 + 3575x_2 + 874x_3 + 210x_4 = 256$

$422x_1 + 543x_2 + 3428x_3 + 428x_4 = 444$

$116x_1 + 256x_2 + 488x_3 + 3627x_4 = 868$

6. SYMMETRICAL EQUATIONS

In a large number of practical problems leading to systems of linear equations it turns out that the determinant of the equations is symmetrical. In the case of symmetrical equations the labor of computing the solution is considerably reduced. For we readily find, by returning to the steps of the elimination, that pairs of elements in (B) which are symmetrically situated with respect to the diagonal are identical except for division by the diagonal element. Hence, both entries can be computed in the course of the same operation on the machine. Suppose an entry below the diagonal has been computed and recorded. While it is still on the dials of the machine we perform the division by the proper diagonal element and record the result in the symmetrical position. Thus all the elements above the diagonal are obtained as a by-product of the process used to obtain those below the diagonal. Of course the elements in the column corresponding to the right-hand sides of the original equations are to be calculated in the usual manner.

7. CHECK ON ACCURACY OF COMPUTATION

The final solution can be checked by substitution back into the original equations. However, it is often more satisfactory to have some sort of current check on the calculation as it proceeds rather than to discover an error after the work is completed.

One rather effective method is to use an additional check column on the right in (A) and (B). Each element in the check column in (A) is the sum of the elements in the same row in (A). The check column in (B) is obtained from the check column in (A) just as is any column to the right of the diagonal, and is calculated along with each row of (B). The check consists in the following relation:

In (B) any element in the check column is equal to one plus the sum of the elements in the same row of (B) to the right of the diagonal.

If we also calculate the set of x's, x_1', x_2', . . ., etc., corresponding to the new right-hand column we have as the final check

$$x_1' = 1 + x_1, \quad x_2' = 1 + x_2, \text{ etc.}$$

For the example of Article 5, the check column and the associated x_j' are as follows:

(A)

11.3923
12.9574
12.5046
14.4656

(B)

1.769678
2.046269
3.577086
0.631813

3.185173	0.439690	3.005319	0.631813
x_1'	x_2'	x_3'	x_4'

The check is seen to be satisfactory, the discrepancies being due to rounding off in the last figure retained.

Exercise

Solve the symmetrical equations and check each step:

$$\begin{aligned}
4.3237x_1 + 1.2124x_2 + 0.3638x_3 - 0.1212x_4 &= 3.8412\\
1.2124x_1 + 6.3660x_2 + 1.2202x_3 - 0.1880x_4 &= 2.3774\\
0.3638x_1 + 1.2202x_2 + 5.2121x_3 + 1.3036x_4 &= 4.2282\\
-0.1212x_1 - 0.1880x_2 + 1.3036x_3 + 4.2277x_4 &= 3.6868
\end{aligned}$$

8. EVALUATION OF DETERMINANTS

The value of the determinant D formed from the coefficients of the given equations is readily calculated from the matrix (B) of Article 5, for we have

(1) $$D = a_{11}\, b_{22}\, c_{33}\, d_{44}$$

This fact is readily verified by reference to the steps of the elimination in Article 4. All the operations performed in deriving equations (5), (6), (7), and (8) from (1), (2), (3), and (4), excepting the division by a_{11}, leave the value of the determinant of the equations unchanged, so that if D_1 denotes the determinant of equations (5) to (8) we have

$$D = a_{11}\, D_1$$

But evidently D_1 is equal to the three rowed determinant of equations (6), (7), (8). Repetition of this reasoning for each stage of the elimination leads to the result (1). The reasoning applies equally well for any number of equations so that in general <u>the value of the determinant is equal to the product of the diagonal elements of (B)</u>.

This shows that the process used to obtain (B) furnishes a convenient method for calculating the value of any numerical determinant.

For instance the first four columns of (A) and (B) in

the numerical example of Article 5 show the numerical work of evaluating the determinant D, which is found to be

$$D = 436.745$$

Exercises

Evaluate the determinants

1.
$$\begin{vmatrix} 37.624 & 26.315 & -11.228 \\ 17.607 & 40.214 & 19.320 \\ -10.588 & 15.364 & 29.897 \end{vmatrix}$$

2.
$$\begin{vmatrix} 641 & 128 & 419 & 305 \\ 321 & 504 & 218 & 416 \\ 101 & 246 & 482 & 115 \\ 226 & 114 & 218 & 582 \end{vmatrix}$$

3.
$$\begin{vmatrix} 6251 & -2832 & 1646 & -782 \\ -1924 & 5845 & -1824 & 1212 \\ 1055 & -2163 & 7684 & -302 \\ -248 & 1234 & -1565 & 5651 \end{vmatrix}$$

9. CALCULATION OF COFACTORS

A simple extension of the process explained in Article 5 enables us to calculate the n^2 cofactors of a determinant D of order n with much less labor than would be required for the direct evaluation of n^2 determinants of order n - 1. The theory of the process may be illustrated by the case of three equations in three unknowns. Let $x_1^{(1)}$, $x_2^{(1)}$, $x_3^{(1)}$ denote the solution of the equations

$$\begin{aligned} a_{11}x_1 + a_{12}x_2 + a_{13}x_3 &= 1 \\ a_{21}x_1 + a_{22}x_2 + a_{23}x_3 &= 0 \\ a_{31}x_1 + a_{32}x_2 + a_{33}x_3 &= 0 \end{aligned}$$

where the determinant D is not zero. Then

$$Dx_1^{(1)} = A_{11}, \quad Dx_2^{(1)} = A_{12}, \quad Dx_3^{(1)} = A_{13},$$

where the **A**'s are the cofactors of the a's. (Cf. Art. 2, (2)). Similarly if $x_1^{(2)}$, $x_2^{(2)}$, $x_3^{(2)}$, is the solution of the above equations with the same left-hand members but with the right-hand members replaced by 0, 1, 0, we have

$$Dx_1^{(2)} = A_{21}, \quad Dx_2^{(2)} = A_{22}, \quad Dx_3^{(2)} = A_{23}, \text{ etc.}$$

Generalizing these results we may now assert that

$$A_{ij} = Dx_j^{(i)}$$

in which $x_j^{(i)}$ ($j = 1, 2, \ldots, n$) denotes the solution of the set of equations with determinant D and right-hand members all zero, except for the ith equation, where it is 1.

The computational set-up may be arranged in matrix form as shown below:

$$\text{(A)} \qquad \begin{matrix} a_{11} & a_{12} & a_{13} & 1 & 0 & 0 \\ a_{21} & a_{22} & a_{23} & 0 & 1 & 0 \\ a_{31} & a_{32} & a_{33} & 0 & 0 & 1 \end{matrix}$$

The matrix (B) is then computed exactly as in Article 5, after which the x's are obtained in turn for <u>each</u> of the n right-hand columns, forming the matrix (C).

$$\text{(E)} \qquad \begin{matrix} x_1^{(1)} & x_2^{(1)} & x_3^{(1)} \\ x_1^{(2)} & x_2^{(2)} & x_3^{(2)} \\ x_1^{(3)} & x_2^{(3)} & x_3^{(3)} \end{matrix}$$

The matrix (E) in the form here written is the conjugate of the inverse matrix of the original equations.

The complete process is shown below for the same

example that was used in Article 5.

(A)	6.4375	2.1849	-3.7474	1.8822	1	0	0	0
	2.1356	5.2101	1.5220	-1.1234	0	1	0	0
	-3.7362	1.4998	7.6421	1.2324	0	0	1	0
	1.8666	-1.1104	1.2460	8.3312	0	0	0	1

(B)	6.4375	0.339402	-0.582120	0.292381	0.15534	0	0	0
	2.1356	4.485273	0.616501	-0.389677	-0.07396	0.22295	0	0
	-3.7362	2.767874	3.760786	0.904963	0.20876	-0.16409	0.26590	0
	1.8666	-1.743928	3.407719	4.022013	-0.28105	0.23570	-0.22529	0.24863

(E)	0.66625	-0.46897	0.46309	-0.28104	Cofactors divided by D.
	-0.47441	0.54746	-0.37739	0.23570	
	0.46743	-0.37741	0.46978	-0.22529	
	-0.28363	0.23560	-0.22500	0.24863	

If we need to solve a number of different sets of equations, all sets having the same left-hand members but different right-hand members, it is advantageous to calculate the matrix (E), which will be the same for all sets. Then the x_j corresponding to any given right-hand members b_1, b_2, b_3, etc. is given by the formula

$$x_j = x_j^{(1)} b_1 + x_j^{(2)} b_2 + \ . \ . \ . + x_j^{(n)} b_n.$$

The matrix (E) is also useful in determining the weights in the method of Least Squares.

Exercises

Compute the cofactors for the exercises of Article 5.

10. MAGNITUDE OF INHERENT ERRORS

Computational errors may be checked by substitution back into the original equations or by use of a check column. But we have to consider another source of inaccuracy which may seriously affect the value of our results even if the computation itself is perfectly correct. When systems of simultaneous equations arise in practical problems, the coefficients usually are not given exactly, either because they have been derived from empirical data or because they have been expressed in decimal form and rounded off to a convenient number of places. It is a matter of considerable importance to know in a given case what degree of accuracy to expect in the solutions. For example, if we know the coefficients accurately, say to four places of decimals, can we be assured that the solutions also are accurate to four places. A simple example shows that such a conclusion may be completely false. The equations

$$\begin{aligned} x - y &= 1 \\ x - 1.00001y &= 0 \end{aligned}$$

have the solution $x = 100{,}001$, $y = 100{,}000$, while the equations

$$\begin{aligned} x - y &= 1 \\ x - 0.99999y &= 0 \end{aligned}$$

have the solution $x = -99{,}999$, $y = -100{,}000$. The coefficients in the two sets of equations differ by at most two units in the fifth decimal place yet the solutions differ by 200,000.

This type of error arises from the very nature of the equations as they are given to us, and cannot be cured by any improvement in the technique of solution. We shall call it the <u>inherent error</u>. To analyze inherent errors, let us consider the set of equations

$$(1) \qquad \sum_{j=1}^{n} a_{ij}x_j = b_i \qquad (i = 1, 2, \ldots, n)$$

which we suppose have a unique solution $x_1, x_2, \ldots, x_n$. Treating the a's, b's, and x's in (1) as all variable and taking differentials we obtain, after transposition,

$$(2) \quad \sum_{j=1}^{n} a_{ij}dx_j = db_i - \sum_{j=1}^{n} x_j da_{ij} \qquad (i = 1, 2, \ldots, n).$$

If the changes db_i and da_{ij} are given, the right-hand members of (2) are known, since the x_j have been obtained from (1). The left-hand members of (2) have the same matrix of coefficients as the original equations (1). Hence the only additional labor required to calculate the changes dx_j as well as the x_j themselves is to annex the right-hand members of (2) as an extra column on the right in the matrix (A), compute the corresponding column of (B), and then solve for the dx_j in the usual manner. If, however, we have already computed the matrix (E), then the dx_j are readily found by the method of equation (2) Article 2. We have thus obtained the infinitesimal changes dx_j in the x_j due to given infinitesimal changes in the a_{ij} and b_i.

In practice, however, the actual changes in the a_{ij} and b_i are not usually known. All that we know in most cases is that da_{ij} and db_i do not exceed some given quantity e, so that $|da_{ij}| \leqq e$, $|db_i| \leqq e$. Then the right-hand members of equations (2) do not exceed a constant k where

$$k = (1 + \sum_{j=1}^{n} |x_j|)e.$$

Hence, our problem is not to solve a set of equations with definitely known right-hand members, but to determine a bound for the greatest possible values of the dx_j for any choice of right-hand members subject only to the limitation that they do not exceed k.

It will be seen that the following procedure leads to the desired result. In the computation scheme of Article 5 annex to the matrix a column with k in each row. Compute the corresponding column of (B) in the usual manner except that all negative signs are replaced by positive signs, that is, always add, regardless of sign. The same applies to the calculation of the matrix (C) giving the dx_j. It is clear that no actual solution for any given right-hand members (not exceeding k) can give results greater than those found in this manner. Hence, we have obtained bounds for the variations in the x_j due to variations not exceeding e in the a_{ij} and b_i.

One caution is to be heeded in interpreting the results of the foregoing procedure. Our use of differentials in deriving equations (2) from (1) implied that products of the type $da_{ij}dx_j$ could be neglected in comparison with the terms retained. If the values of the dx_j turn out to be too large for this assumption to hold, the actual errors might be even larger than indicated.

Two suggestions may be offered regarding computation.

1) Use 1 instead of k in the annexed column in (A), and multiply the final solution by k. The result is the same, and the work more convenient, especially if we want to consider several different choices of k.

2) Since the purpose of finding the dx_j is to place a bound on the error, precise accuracy in computing the dx_j is unnecessary. If two or three significant figures are used throughout, the resulting values will be sufficiently accurate to indicate the limits within which the error must lie. Since we need use only two or three figures in the computation and since we do not have to watch signs, the computation for the errors is quite simple.

Example 1. If the coefficients of the numerical example of Article 5 are known to be in error less than five units in the fifth decimal place, what is the maximum error in the solution.

Here we have e = 0.00005. By equation (3) and the values of the x_j found in Article 5 we have

$$k \leqq (1 + 2.2 + 0.56 + 2.0 + 0.37)e < 0.00031$$

Next annex a column of 1's on the right of (A) and compute

the corresponding column in (B) obtaining

(B)

0.16
0.30
0.65
1.00

The corresponding one-rowed matrix (C) is

1.92	1.64	1.55	1.00

These values might appropriately be called <u>measures of sensitivity</u>, since they give bounds for the changes in the x_j due to unit changes in the right-hand members of the equations. When these are multiplied by k we have $dx_1 = 0.00059$, $dx_2 = 0.00051$, $dx_3 = 0.00048$, $dx_4 = 0.00031$. These bounds give us an idea of the attainable accuracy in the solution of the given example when the coefficients are accurate only to four places.

In calculating these bounds we assumed the worst possible combination of errors in the coefficients, and since such an event is highly improbable, the actual errors may be considerably less than the bounds indicate. We see however that it is not safe to assume that the solutions are reliable in the fourth decimal place even though our <u>computational</u> check indicated a maximum error of three units in the sixth place.

<u>Example 2</u>. The following example shows the complete computation for the solution of a set of equations, with a check column as a control on the accuracy of the <u>computation</u>, and a check column to determine the inherent errors of the solution assuming that the original coefficients are accurate to four decimal places.

x_1	x_2	x_3	x_4		Computation Check Column	Inherent Error Column
6.4375	2.9042	-7.1313	5.8024	2.1345	10.1473	1
2.1356	1.0124	-2.3121	1.9011	1.3161	4.0531	1
-3.7362	-1.6421	4.0526	-3.3515	1.8422	-2.8350	1
1.8666	0.8526	-2.0041	1.6824	2.4430	4.8405	1
6.4375	0.45113786	-1.10777476	0.90134369	0.33157282	1.5727961	.155
2.1356	0.04894999	1.09629802	-0.48640631	12.42069887	14.03059050	27.2
-3.7362	0.04344127	-0.13389264	-0.27806132	-18.98126317	-18.25932484	20.5
1.8666	0.01050607	0.05215458	0.01956426	137.1661004	138.1661014	135.

-128.3048324	58.1348268	19.1593237	137.1661004	Solution
-127.3048336	59.1348274	20.1593237	138.1661014	Computation Check
257	157	58.3	135	Measures of Sensitivity

Here $e = 0.00005$

Hence $k = [1 + 128 + 58 + 19 + 137]\ (0.00005) = 0.0172$

Hence

$dx_1 = 4.4 \quad dx_2 = 2.7 \quad dx_3 = 1.0 \quad dx_4 = 2.3$ Bounds for Inherent Errors

This example illustrates vividly the important difference between a check on the accuracy of the numerical computation and a check of the attainable accuracy when the coefficients are given as decimal approximations. We could increase the accuracy of the computation by carrying a greater number of places throughout the work. In this case our results, as far as the computation is concerned, are accurate to five decimal places. But the attainable accuracy with the given coefficients is not under our control, being due to the character of the equations themselves. We see that actually the first digit to the left of the decimal point is not reliable, and that obtaining the solution to seven places of decimals was a sheer waste of labor.

Exercises

Determine bounds for the inherent errors in the solution of the exercises of Articles 5 and 7, assuming that the coefficients are accurate only to the given number of places.

There is still another source of error which becomes increasingly significant as the number of equations increases. Suppose that the computational work is carried to n places of decimals. If the number of equations is large, the elimination will involve many thousand arithmetical operations, and the cumulative effect of errors due to rounding off at n places may eventually be so great as to make the final results completely worthless. An analysis of this problem is too complex for inclusion here. The reader may consult the article by John von Neumann and H. H. Goldstine, Bull. Amer. Math. Soc., Vol. 53, p. 1021, November, 1947.

Chapter II

SOLUTION OF EQUATIONS BY SUCCESSIVE APPROXIMATIONS

It is usually impractical to solve either transcendental equations or algebraic equations of higher than the second degree by means of direct analytical operations.

Practically all of the many methods which have been devised to solve such equations are in effect methods of approximation whereby a crude guess at the root is used to obtain a closer value, the latter again used to obtain a still closer value, and so on, until the desired accuracy is secured. Some of the most useful ways of carrying out the method of successive approximations are described and illustrated in this chapter.

11. ONE EQUATION IN ONE UNKNOWN

Suppose that the equation to be solved is

$$y = f(x) = 0. \tag{1}$$

Suppose also that by means of a rough graph or otherwise we have ascertained that there is a root of the equation in the vicinity of $x = x_0$. The method of successive approximations consists in finding a sequence of numbers $x_0, x_1, x_2, \ldots,$ converging to a limit a such that $f(a) = 0$. The recurrence relation by which x_{n+1} is calculated after x_n has been obtained may be expressed in the form

$$x_{n+1} = x_n - f(x_n)/m, \tag{2}$$

in which m denotes the slope of a suitably chosen line. The ideal choice of m is obviously the slope of the chord joining the point (x_n, y_n), where $y_n = f(x_n)$, to the point (a, 0), for then $x_{n+1} = a$, and the problem is solved. Since of course the point a is unknown this ideal value of m is also unknown and we are obliged to use some type of approximation for m. There are several ways of choosing an approximate value for m.

1) The slope of the tangent to the curve $y = f(x)$ at $x = x_n$ gives

$$m = f'(x_n) \qquad \left(f'(x) = \frac{df(x)}{dx}\right)$$

2) The slope of the chord joining two points already calculated, say (x_i, y_i) and (x_n, y_n), gives

$$m = \frac{y_n - y_i}{x_n - x_i}$$

3) If $x = x_1$ and $x = x_2$ are values of x for which $f(x_1)$ and $f(x_2)$ have unlike signs, so that the desired root lies between x_1 and x_2, it is frequently satisfactory to use

$$m = \frac{y_1 - y_2}{x_1 - x_2}$$

throughout the successive steps. Since m is calculated once for all, this choice saves considerable labor.

4) Similarly if the curvature is not too great near the root it may suffice to use

$$m = f'(x_1)$$

throughout the successive steps. This is especially advantageous if the computation of successive values of $f'(x)$ is laborious.

5) If the curvature does not change sign near the root, it is clear from a consideration of the graph that a value of m between those given by 3) and 4) will often

be better than either one. Hence, we may employ the arithmetic mean

$$m = 1/2 \left[f'(x_1) + \frac{y_1 - y_2}{x_1 - x_2} \right]$$

as our approximate value of m.

No inflexible rule can be given for the best choice of m. The computer need only remember that the ideal value is the slope of the chord joining a known point (x_1, y_1) on the curve with the point (a, 0), and then make the wisest choice available in the particular problem. The several methods mentioned above will now be illustrated by numerical examples.

Example 1. Find the positive root of

$$f(x) = x^3 - x - 4 = 0.$$

Here we have, using method 1),

$$x_{n+1} = x_n - \frac{f(x_n)}{f'(x_n)} = \frac{2x_n^3 + 4}{3x_n^2 - 1}$$

By substituting a few trial values of x we find that f(1) = -4, f(2) = 2, indicating a root between x = 1 and x = 2, probably nearer x = 2. Hence x_0 is chosen as $x_0 = 2$. The computation may be arranged as shown below:

n	x_n	$2x_n^3 + 4$	$3x_n^2 - 1$	x_{n+1}
0	2	20	11	1.8
1	1.8	15.664	8.72	1.7963
2	1.7963	15.59222	8.680082	1.79632
3	1.79632	15.592605	8.680259	1.79632

The desired root to 5 decimal places is x = 1.79632.

Example 2. Find the smaller positive root of

$$x^{1.8632} - 5.2171x + 2.1167 = 0.$$

Here $f(0) = 2.1$, $f(1) = -2.1$, approximately. Using method 3), we take $m = -4.2$ and $x_0 = 0.5$. The computation follows:

n	x_n	$\log x_n$	$\log x_n^{1.8632}$	$x_n^{1.8632}$	$f(x_n)$	m	x_{n+1}
0	0.5	-0.30103	9.43912 - 10	0.27487	-0.21698	-4.2	0.4484
1	0.4484	-0.34833	9.35099 - 10	0.22438	0.00173	-4.2	0.44881
2.	0.44881	-0.34791	9.35177 - 10	0.22478	0		

To five places the desired root is $x = 0.44881$.

Example 3. Find the positive root of

$$f(x) = x - 2 \sin x.$$

By method 1)

$$x_{n+1} = x_n - f(x_n)/f'(x_n) = \frac{-2x_n \cos x_n + 2 \sin x_n}{1 - 2 \cos x_n} = \frac{N}{D}$$

From a rough sketch we conclude that the root is somewhere near $x = 2$. The computation may be arranged as shown:

n	x_n	N_n	D_n	x_{n+1}
0	2	3.48	1.832	1.90
1	1.90	3.1211	1.6466	1.8955
2	1.8955	3.1049258	1.6380560	1.8954942
3	1.8954942	3.104904759	1.638044922	1.895494267

The last number in the right-hand column is the root to nine places.

Example 4. Find the largest positive root of

$$x^4 - 2.0379x^3 - 15.4245x^2 + 15.6696x + 35.4936 = 0$$

By trial substitutions we locate the largest positive root between $x = 4$ and $x = 5$. Since $f(4) = -23$, $f(5) = 94$, approximately, the slope of the chord is 117. The approximate value of $f'(4)$ is 50. Using method 5) we take a value of m about half way between 50 and 117, say $m = 85$, and let $x_0 = 4$. The substitutions are performed by the usual method of synthetic substitution. This can be done in a continuous operation on the calculating machine. (In the case of machines where the carry-over is not effective over all the dials in certain positions it may be necessary to replace negative numbers on the dials by their complements, in which case the change of sign must be carefully watched.) The work may be arranged as shown:

n	x	f(x)	f(x)/m	m
0	4	-23	-0.27	85
1	4.27	-5.05	-0.059	85
2	4.329	+0.137	+0.00162	85
3	4.32738	-0.01156	-0.000136	85
4	4.327516	+0.000924	+0.0000109	85
5	4.327505			

It is possible to prove that with suitable limitations on $f(x)$ and on the choice of the initial value x_0 the process of successive approximations will give a sequence x_0, x_1, x_2, ... which converges to a root of $f(x) = 0$. For the practical computer, however, such a theorem is of somewhat academic interest, since the numerical process itself either converges with reasonable rapidity to a value which is obviously a root, or else by its behavior gives warning that x_0 was poorly chosen or that $f(x)$ has some peculiarity near the supposed root. Some peculiar cases will be considered in the next article.

Exercises

Find to five places all real roots of

1. $x^4 - 2x^3 - 4x^2 - 4x + 4 = 0$
2. $x^{\sqrt{2}} + \sqrt{2}x = 6$
3. $2 \cos x - x^2 = 0$
4. $4x^4 - 24x^3 + 44x^2 - 24x + 3 = 0$
5. $2x^2 + 2 \sin^2 x = 5$

12. EXCEPTIONAL CASES

If the derivative $f'(x)$ vanishes at or near a root of $f(x) = 0$, the process of approximation encounters trouble because the divisor m is small. In such a case it is frequently best to obtain the root of $f'(x) = 0$ first of all, especially if $f''(x)$ is not near zero. Suppose that we have found the root $x = a$ of the equation $f'(x) = 0$. We next calculate $f(a)$ and $f''(a)$. Then

1) If $f(a) = 0$ the quantity a is a double root.

2) If $f(a) \neq 0$ and $f(a)$ and $f''(a)$ have like signs, there is no root of $f(x) = 0$ in the vicinity of $x = a$. For under these conditions the curve $y = f(x)$ is concave away from the x-axis and cannot cross the axis in the neighborhood of $x = a$.

3) If $f(a) \neq 0$ and $f(a)$ and $f''(a)$ have unlike signs, we may expect to find two roots of $f(x) = 0$, one greater than a and one less than a. Using Taylor's series for $f(x)$ at $x = a$, noting that $f'(a) = 0$ and neglecting terms of third and higher degree we obtain the approximate values

$$x_0' = a + \sqrt{-f(a)/\tfrac{1}{2}f''(a)}, \tag{1}$$

$$x_0'' = a - \sqrt{-f(a)/\tfrac{1}{2}f''(a)}, \tag{2}$$

for the two roots. Each one of these may now be refined by successive approximations in the usual manner.

Example 1.

$$f(x) = 2x^4 + 16x^3 + x^2 - 74x + 56 = 0$$

Tabulating a few values of f(x) and its derivatives we have

x	f(x)	f'(x)	f"(x)
0	56	-74	2
1	1	-16	122
2	72	186	290

These values show a root of $f'(x) = 0$ between $x = 1$ and $x = 2$, with the possibility of two roots of $f(x) = 0$ in the same interval. The method of approximation applied to $f'(x) = 0$ gives

n	x_n	$f'(x_n)$	$f''(x_n)$
0	1	-16	122
1	1.13	1.0944	141
2	1.12224	0.00376	140
3	1.1222132	0.0000102	140
4	a = 1.122213127		

We next find $f(a) = -0.00005576$, $\frac{1}{2}f''(a) = 69.98$, and these values in conjunction with formulas (1) and (2) above yield as a first approximation for the two roots of $f(x) = 0$

$$x_0' = 1.12310582, \quad x_0'' = 1.12132044.$$

For the first value x_0'

$$f(x_0') = 0.000000024292, \quad f'(x_0') = 0.12500$$

and the method of approximation gives the improved value

$$x_1' = 1.1231056256$$

Similarly

$$x_1'' = 1.1213203436.$$

Example 2.

$$f(x) = 2 + 3x - 6 \sin x = 0.$$

Here $f'(x) = 3 - 6 \cos x$, which vanishes for $a = \pi/3 = 1.047197551$. Also $f(a) = -0.05456$, $1/2f''(a) = 2.598$. Use of formula (1) gives as an approximate value for the root

$$x_0' = 1.192111$$

The method of approximation applied to this value now yields

x	f(x)	f'(x)
1.192111	0.001427	0.7818
1.190286	0.000006662	0.7716
1.190277366	0.000000002	0.77
1.190277364		

The other root is found in a similar manner.

When all three quantities, $f(x)$, $f'(x)$, $f''(x)$ vanish close to the same point additional complications occur. As such instances are rare and the analysis is lengthy these cases will be omitted.

Exercises

Find to five places the roots of

1. $x^3 - 6.0266x^2 + 4.3048x + 15.9533 = 0.$
2. $x^4 + 12x^3 - 9.5x^2 - 6x + 4.5 = 0$
3. $x^4 - 6x^3 - 113x^2 + 504x + 2436 = 0.$
4. $x^4 + 16x^3 + 11x^2 - 224x + 286 = 0.$
5. $2 + 7\sqrt{2x} - 14 \sin x = 0.$

13. SIMULTANEOUS EQUATIONS

The method of approximation may also be extended to the solution of simultaneous equations. Thus if

$$F(x, y) = 0, \quad G(x, y) = 0,$$

are two equations in two unknowns, and if a point (x_1, y_1) close to a solution has been determined by graphical methods or otherwise a closer point (x, y) can usually be obtained as follows:

Let $x - x_1 = \delta x$, $y - y_1 = \delta y$. Expand $F(x, y)$ and $G(x, y)$ in Taylor's series to terms of the first degree, and assume that (x, y) is a solution, i. e., $F(x, y) = G(x, y) = 0$. Then, approximately,

$$0 = F + F_x \delta x + F_y \delta y,$$
$$0 = G + G_x \delta x + G_y \delta y,$$

in which for brevity we have set $F = F(x_1, y_1)$, $F_x = \frac{\partial F}{\partial x}$ at $x = x_1$, $y = y_1$, etc. The two equations are solved for δx and δy and the new approximation to the solution is given by

$$x = x_1 + \delta x, \quad y = y_1 + \delta y.$$

The process is repeated until the desired accuracy is secured.

The following scheme of computation is convenient for the case of two equations.

F	F_x	F_y
G	G_x	G_y
D	$-\delta x$	$+\delta y$
	x	y

The values of F, G, F_x etc. are calculated for $x = x_1$, $y = y_1$ and inserted in the proper places in the scheme. Then the first column is covered and the remaining two-rowed determinant

$$D = F_x G_y - F_y G_x$$

is evaluated on the machine by cross multiplication and entered in the place shown. Next the second column is covered and the remaining two-rowed determinant

$$FG_y - GF_y$$

is evaluated on the machine, and while the result is still on the dials the division by D is performed. This gives $-\delta x$, which is entered as shown. Finally the third column is covered, the determinant $FG_x - GF_x$ is evaluated, the result is divided by D, and recorded for δy. The improved values of x and y are now given by $x = x_1 + \delta x$, $y = y_1 + \delta y$ and are recorded. The next step will be an exact repetition of the foregoing with the new values of x and y.

If it is found that the values of F_x, F_y, G_x, G_y are not much changed in successive calculations, we need not recompute them at every step but merely copy them, together with D, from the previous step.

Example. Find a solution (different from the obvious solution $x = 0$, $y = 0$) of the equations

$$x - \sin x \cosh y = 0,$$
$$y - \cos x \sinh y = 0.$$

An examination shows that in the first quadrant a solution occurs in each interval in which both $\sin x$ and $\cos x$ are positive, i. e., for $2\pi < x < \frac{5\pi}{2}$, $4\pi < x < \frac{9\pi}{2}$, etc. Accordingly, we first make the change of variable $x = \frac{5\pi}{2} - x'$ which transforms the given equation to

$$\frac{5\pi}{2} - x' - \cos x' \cosh y = 0,$$
$$y - \sin x' \sinh y = 0.$$

We take as initial estimates for x' and y the values $x' = 0.2$, $y = 2.9$, and carry out the computation according to the scheme described above.

	$x' = 0.2$	$y = 2.9$
-1.279 1.100 -79.49	0.8107 -8.879 -0.136 $x' = 0.336$	-8.879 -0.8107 -0.132 $y = 2.768$
-0.029562 0.15277 -58.751	1.6359 -7.4883 -0.02029 $x' = 0.35629$	-7.4883 -1.6359 0.00049 $y = 2.76849$
0.001496 0.0004830 -58.51	1.790 -7.437 -0.00002 $x' = 0.35631$	-7.437 -1.790 0.00020 $y = 2.76869$
0.0000563 0.0000455 -58.5*	1.79* -7.44* -0.00000 $x' = 0.35631$	-7.44* -1.79* 0.00001 $y = 2.76870$
*Not recomputed.	$x = 7.85398 - x'$ $x = 7.4977$	$y = 2.7687$

14. SUCCESSIVE SUBSTITUTIONS

It may be possible by suitable manipulation and combination of the given equations $F(x, y) = 0$, $G(x, y) = 0$ to transform them to an equivalent pair

$$x = f(x, y)$$
$$y = g(x, y)$$

such that for values of x and y near a common solution the following inequalities are satisfied:

$$\left|\frac{\partial f}{\partial x}\right| + \left|\frac{\partial f}{\partial y}\right| < k, \qquad \left|\frac{\partial g}{\partial x}\right| + \left|\frac{\partial g}{\partial y}\right| < k,$$

where k denotes some positive constant less than unity. Whenever this is possible a process of successive substitutions will lead to a solution.

Example. The pair of equations used in the example of Article 13 can be expressed in the form

$$y = \cosh^{-1}\left(\frac{\frac{5\pi}{2} - x'}{\cos x'}\right), \qquad x' = \sin^{-1}\left(\frac{y}{\sinh y}\right).$$

We take $x' = 0$ in the first equation, compute y, put this y in the second equation and compute x', put this x' in the first equation and compute y, etc. This process gives us the following sequence of values:

x'	y
0	2.75
0.36	2.769
0.35614	2.76865
0.35631	2.76868

These values check closely with the values obtained for x' and y in the Example of Article 13.

Exercises

Solve by the methods of Articles 13 or 14.

1. $x^3 - 3xy^2 = 4,$
   $3x^2y - y^3 = 2.$

2. $2x = \sin 1/2(x - y).$
   $2y = \cos 1/2(x + y).$

3. $2 \cos x \cos y = 1,$
   $\sin y = 2 \sin x$

4. $x^{1.46} + y^{2.38} = 12,$
   $x^{2.38} - y^{1.46} = 5.$

5. $x^{10} + y^{10} = 1024,$
   $e^x - e^y = 1.$

15. EXCEPTIONAL CASES

When the determinant

$$D = \begin{vmatrix} F_x & F_y \\ G_x & G_y \end{vmatrix}$$

vanishes at or near a supposed solution of $F(x, y) = 0$, $G(x, y) = 0$ we may anticipate difficulty with the method described in Article 13. The vanishing of D indicates (a) multiple solutions, (b) two or more solutions close together, or (c) no solution at all. The locus of points defined by the equation $D(x, y) = 0$ is the curve on which the loci of $F(x, y) = \text{const.}$, $G(x, y) = \text{const.}$ have either common tangents or singular points. If the two determinants

$$(DF) = \begin{vmatrix} D_x & D_y \\ F_x & F_y \end{vmatrix}, \qquad (DG) = \begin{vmatrix} D_x & D_y \\ G_x & G_y \end{vmatrix}$$

are not both zero in the vicinity of the point in question we may proceed as follows:

I. Construct a graph of the curves showing the approximate point of intersection or tangency, and also showing the signs of the functions $F(x, y)$ and $G(x, y)$ in the several regions bounded by the curves.

II. Solve the simultaneous set $F = 0$, $D = 0$ (or $G = 0$, $D = 0$, whichever seems easier) by successive approximations. We suppose that the intersection of $F = 0$, $D = 0$ is found to be $x = a$, $y = b$. Then calculate $G(a,b)$. If $G(a, b) = 0$, the curves are tangent at (a, b) and the point (a, b) is the desired solution. If $G(a, b) \neq 0$ its sign, compared with the signs of F and G as shown in I, will determine whether

a) There is no solution, or

b) Two nearby solutions.

III. If case b) occurs we expand $F(x, y)$ and $G(x, y)$ in Taylor's series at (a,b), assume $F(x,y) = 0$, $G(x, y) = 0$; recall that $F(a, b) = 0$, and obtain

$$(1) \qquad 0 = 0 + F_x\,\delta x + F_y\,\delta y + 1/2F_{xx}\,\delta x^2 + F_{xy}\,\delta x\,\delta y + 1/2F_{yy}\,\delta y^2,$$

$$(2) \qquad 0 = G + G_x\,\delta x + G_y\,\delta y + 1/2G_{xx}\,\delta x^2 + G_{xy}\delta x\delta y + 1/2G_{yy}\,\delta y^2,$$

where $\delta x = x - a$, $\delta y = y - b$.

The solution of this pair of simultaneous quadratic equations for δx and δy normally would be a considerable task in itself. Here however we are aided by two facts.

1. Since $D = F_xG_y - F_yG_x = 0$ we have

$$\frac{G_x}{F_x} = \frac{G_y}{F_y} = k$$

so that if we multiply (1) by k and subtract from (2) we eliminate all terms of first degree and obtain

$$(3) \qquad 0 = G + A\,\delta x^2 + B\,\delta x\,\delta y + C\,\delta y^2$$

in which $A = 1/2(G_{xx} - kF_{xx})$, etc.

2. If we set $\delta y = \mu\,\delta x$ we may expect that μ will differ only slightly from $m = -F_x/F_y$, the slope of the tangent to $F(x, y) = 0$ at (a, b).

Therefore setting $\delta y = \mu\,\delta x$ we rewrite equation (1) in the form

$$(4) \qquad \mu = -\frac{F_x}{F_y} - \frac{\delta x}{2F_y}\left[F_{xx} + 2\mu F_{xy} + \mu^2 F_{yy}\right],$$

and equation (3) in the form

(5) $$\mu x = \pm \sqrt{\frac{-G}{A + \mu B + \mu^2 C}}.$$

(If F_y is small compared to F_x it is better to set $\delta x = \nu \delta y$ and solve for ν and δy.) The two equations (4) and (5) are easily solved by successive substitutions. We then have approximations to the two points of intersection (x_1, y_1), (x_2, y_2).

IV. Each of these trial pairs is now refined to the desired degree of accuracy by the usual method of successive approximations.

Example. Investigate the solutions of

$$F = x^4 + y^4 - 67 = 0,$$

$$G = x^3 - 3xy^2 + 35 = 0.$$

The stages of the investigation as outlined above are as follows.

I. The graph of the curves, showing the signs of F and G, is given in Fig. 1 and indicates points of tangency at approximately (1.9, $\pm$ 2.7).

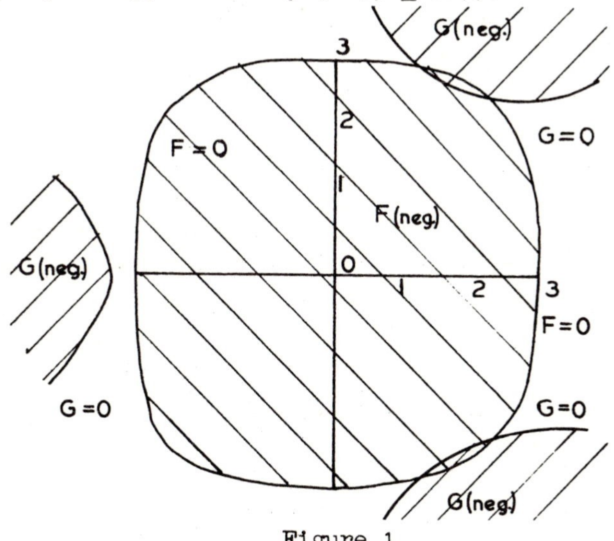

Figure 1

II. The general expression for D as a function of x and y proves to be

$$D = 12y \left[y^4 - x^2y^2 - 2x^4 \right]$$

Accordingly, we solve the simultaneous equations

$$x^4 + y^4 - 67 = 0,$$
$$y^4 - x^2y^2 - 2x^4 = 0.$$

(It happens here that these equations can be solved analytically, but usually we must resort to the method of approximations.) The solution in the first quadrant is

$$a = 1.913269726, \qquad b = 2.705771994.$$

For these values $G(a, b) = -0.018586$, and the fact that $G(a, b)$ is negative shows, by a consideration of the signs in Fig. 1, that there are two nearby points of intersection.

III. At the point (a, b) we now find

$$F_x = 20.015, \quad F_y = 79.238, \quad m = -0.35356$$
$$G_x = -10.982, \quad G_y = -31.061, \quad k = -0.39200$$

$$1/2F_{xx} = 21.96 \quad F_{xy} = 0 \quad 1/2F_{yy} = 43.93$$
$$1/2G_{xx} = 5.740 \quad G_{xy} = -16.23 \quad 1/2G_{yy} = -5.740$$

Using these values we set up equations (4) and (5) as in the text. They prove to be

$$(4) \qquad \mu = -0.35356 - (0.277 + 0.554\,\mu^2)\,\delta x$$

$$(5) \qquad \delta x = \pm \sqrt{\frac{0.018586}{14.35 - 16.23\,\mu + 11.48\,\mu^2}}.$$

Starting with the trial value $\mu = -0.35356$ in (5) and

using successive substitutions we get (choosing the negative sign)

$$\mu = -0.34337, \qquad \delta x = -0.02956, \qquad \delta y = 0.01015$$

This gives $x = 1.88371$, $y = 2.71592$.

IV. Returning to the original equations $F = 0$ and $G = 0$, we solve by successive approximations, starting with the values $x = 1.88371$, $y = 2.71592$, and after three steps obtain

$$x = 1.88364521, \qquad y = 2.71594754.$$

The other solution is found in a similar manner.

Exercises

Solve the sets of equations.

1. $$13x^2 - 8xy + 5y^2 - 54x + 22y + 54 = 0,$$
$$5x^2 - 16xy + 3y^2 - 42x + 42y + 56 = 0.$$

2. $$2x^2 - xy - 3y^2 + 7x - 18y + 9 = 0,$$
$$3x^2 - 14xy + 8y^2 + 10x - 15y + 5 = 0.$$

16. COMPLEX ROOTS OF ALGEBRAIC EQUATIONS

Corresponding to a pair of complex roots $x \pm iy$ of the algebraic equation

$$f(z) = a_0 z^n + a_1 z^{n-1} + \dots a_n = 0$$

with real coefficients there is a real quadratic factor $z^2 + pz + q$. Instead of determining the complex root directly by the method of Article 11, we may avoid all substitutions of complex numbers by determining the p and q of the corresponding quadratic factor. To do this in a systematic manner with the calculating machine we set up a procedure for synthetic division by a quadratic factor analogous to the well known synthetic division by a linear factor. The coefficients a_0, a_1, ..., a_n are arranged in a column. Then a second column of b's is computed by the formulas

$$b_i = a_i - pb_{i-1} - qb_{i-2}.$$

The b's are the coefficients in the quotient. In the same way we divide the quotient by $z^2 + pz + q$, obtaining a column of c's. The complete computational setup is shown below:

a_0	b_0	c_0
a_1	b_1	c_1
a_2	b_2	c_2
.	.	.
.	.	.
.	.	.
a_{n-4}	b_{n-4}	c_{n-4}
a_{n-3}	b_{n-3}	c_{n-3}
a_{n-2}	b_{n-2}	c_{n-2}
a_{n-1}	b_{n-1}	c_{n-1}
a_n	b_n	
b_{n-1}	c_{n-2}	c_{n-3}
b_n	c_{n-1}	c_{n-2}
D	δp	$-\delta q$
	p	q

The recursion formulas are

$$b_i = a_i - pb_{i-1} - qb_{i-2}, \qquad (i = 0, 1, \ldots n),$$

$$c_i = b_i - pc_{i-1} - qc_{i-2}, \qquad (i = 0, 1, \ldots, n - 2),$$

$$c_{n-1} = \quad - pc_{n-2} - qc_{n-3}.$$

Also $D = c_{n-2}^2 - c_{n-1}c_{n-3}$,

$$\delta p = (b_{n-1}c_{n-2} - b_n c_{n-3})/D,$$

$$-\delta q = (b_{n-1}c_{n-1} - b_n c_{n-2})/D.$$

The quantities D, δp, and $-\delta q$ are computed on the calculating machine just as were the D, $-\delta x$, and δy in the scheme shown on page 45.

After a few steps it will generally be unnecessary to recompute the c's and the D, thus saving considerable labor.

In case the determinant D is zero or nearly zero we may conclude that the given equation has two or more pairs of equal roots, or two or more pairs nearly equal. (This conclusion of course holds only if b_{n-1} and b_n are also small.)

Example. Solve the equation

$$z^4 - 3z^3 + 20z^2 + 44z + 54 = 0.$$

This equation has no real roots. Without going to the trouble of hunting for the approximate location of its complex roots let us simply start with the trial values $p = 0$, $q = 0$ and see what happens. The computation follows. The convergence is slow at the start because $p = 0$, $q = 0$ were not very close to the actual values of p and q.

Step 1.	$p_0 = 0$	$q_0 = 0$
1	1	1
-3	-3	-3
20	20	20
44	44	0
54	54	
44	20	-3
54	0	20
400	2.6	-2.7

Step 2.	$p_1 = 2.6$	$q_1 = 2.7$
1	1	1
-3	-5.6	-8.2
20	31.86	50.48
44	-23.716	-109.11
54	29.640	
-23.716	50.48	-8.2
29.640	-109.11	50.48
1654	-0.58	(-) 0.66
Step 3.	$p_2 = 2.02$	$q_2 = 2.04$
1	1	1
-3	-5.02	-7.04
20	28.100	40.28
44	-2.541	-67.00
54	1.809	
-2.541	40.28	-7.04
1.809	-67.00	40.28
1151	-0.078	(-) 0.085
Step 4.	$p_3 = 1.942$	$q_3 = 1.955$
1	1	1
-3	-4.942	-6.884
20	27.642	39.055
44	-0.1915	-62.39
54	-0.00292	
-0.01915	39.055	-6.884
-0.00292	-62.39	39.055
1096	-0.00070	(-) 0.00119

Step 5.	$p_4 = 1.94130$	$q_4 = 1.95381$
1	1	
-3	-4.94130	
20	27.6387	
44	-0.000647	
54	+0.000487	
1096	-0.0000200	(-) 0.0000195
Step 6.	$p_5 = 1.941280$	$q_5 = 1.953791$
1	1	
-3	-4.941280	
20	27.638617	
44	-0.000066	
54	+0.000047	
1096	-0.0000020	(-) 0.0000019
	$p_6 = 1.941278$	$q_6 = 1.953789$

$$z = -0.970639 \pm 1.005808i.$$

The remaining roots are now easily found from the equation

$$z^2 - 4.941280z + 27.638617 = 0$$

Exercises

Find the complex roots of

1. $x^6 - 2x^5 + 2x^4 + x^3 + 6x^2 - 6x + 8 = 0$
2. $x^4 + 3x^3 - 3x^2 + 23x + 180 = 0$

17. SOLUTION OF λ-DETERMINANTS

In Article 3, we considered briefly the problem of finding non-zero solutions in $x_1, x_2, \ldots, x_n$ of linear equations in the form

(1)
$$\sum_{j=1}^{n} a_{ij}x_j = \lambda x_i \qquad i = 1, 2, \ldots, n,$$

where the values of λ are to be determined for which non-zero solutions exist. These values of λ are roots of the n-th degree algebraic equation

(2)
$$\begin{vmatrix} a_{11}-\lambda & a_{12} & \ldots & a_{1n} \\ a_{21} & a_{22}-\lambda & \ldots & a_{2n} \\ \cdot & \cdot & \cdot & \cdot \\ a_{n1} & a_{n2} & \ldots & a_{nn}-\lambda \end{vmatrix} = 0$$

and corresponding to each root λ_k there exists a set of x's, $x_1^{(k)}, x_2^{(k)}, \ldots, x_n^{(k)}$, not all zero, satisfying equations (1) for $\lambda = \lambda_k$. In all important practical applications in which this problem arises the determinant in (2) is symmetrical and moreover the coefficients a_{ij} are such that the expression

$$\sum_{j=1}^{n} \sum_{j=1}^{n} a_{ij}x_ix_j$$

is a positive definite quadratic form. In this case the method of successive approximations can be used not only to find the numerical value of the roots λ_k but also at the same time to give the corresponding $x_1^{(k)}$, $x_2^{(k)}, \ldots, x_n^{(k)}$.

Under the given hypotheses it can be proved that

1) all the roots λ_k of equation (2) are real and positive;

2) any two sets of x's, $x_1^{(k)}, x_2^{(k)}, \ldots, x_n^{(k)}, x_1^{(q)}$, $x_2^{(q)}, \ldots, x_n^{(q)}$, corresponding respectively to two different roots λ_k and λ_q satisfy the condition of orthogonality; i. e.,

3) $x_1^{(k)} x_1^{(q)} + x_2^{(k)} x_2^{(q)} + \dots + x_n^{(k)} x_n^{(q)} = 0.$

Furthermore, since the equations (1) are homogeneous it is evident that if $x_1^{(k)}, x_2^{(k)}, \dots, x_n^{(k)}$ is a solution of (1) corresponding to $\lambda = \lambda_k$, then $cx_1^{(k)}, cx_2^{(k)}, \dots, cx_n^{(k)}$, where c is any constant, is also a solution. Hence one of the x's, say x_n, may be chosen arbitrarily and the values of the other x's will then in general be uniquely determined.

To show how the method of successive approximations works we apply it to the numerical example

$$\lambda x_1 = 4x_1 + 2x_2 + 2x_3,$$
$$\lambda x_2 = 2x_1 + 5x_2 + x_3,$$
$$\lambda x_3 = 2x_1 + x_2 + 6x_3.$$

First let $x_3 = 1$, and put the three equations in the form

$$(4) \qquad \begin{cases} x_1 = \frac{1}{\lambda}(4x_1 + 2x_2 + 2) \\ x_2 = \frac{1}{\lambda}(2x_1 + 5x_2 + 1) \end{cases}$$

$$(5) \qquad \lambda = 2x_1 + x_2 + 6$$

Next choose any convenient trial values for x_1 and x_2, say $x_1 = x_2 = 1$. From (5) we now get $\lambda = 9$. Putting the trial values of x_1, x_2, and λ on the right in (4) we find new values $x_1 = 0.89$, $x_2 = 0.89$. By repeated substitutions in this manner we obtain the convergent sequence

x_1	x_2	x_3	λ
1	1	1	9
0.89	0.89	1	8.67
0.85	0.83	1	8.53
0.83	0.80	1	8.46
0.81	0.78	1	8.40
0.805	0.770	1	8.38
0.806	0.771	1	8.383
0.807	0.771	1	8.385
0.8074	0.7715	1	8.3863
0.8076	0.7717	1	8.3869
0.8076	0.7719	1	8.3871
0.8077	0.7720	1	8.3874

We now have one value of λ, $\lambda_1 = 8.3874$, and the corresponding values of x_1, x_2, x_3. To proceed we make use of the orthogonality relationship (3). For if λ_2 is another root and $x_1^{(2)}$, $x_2^{(2)}$, $x_3^{(2)}$ are the corresponding x's then by (3)

$$0.8077x_1^{(2)} + 0.7720x_2^{(2)} + x_3^{(2)} = 0.$$

Using this equation we eliminate one x, say x_3 from the original equations, set $x_2 = 1$, and put the equations in the form

$$x_1 = \frac{1}{\lambda}(2.3846x_1 + 0.4560)$$

$$\lambda = 1.1923x_1 + 4.2280$$

The process of successive substitutions now gives

x_1	x_2	λ
1	1	5.42
0.52	1	4.85
0.35	1	4.64
0.28	1	4.56
0.25	1	4.53
0.23	1	4.50
0.223	1	4.494
0.220	1	4.490
0.218	1	4.488
0.2174	1	4.487
0.2171	1	4.4868
0.2170	1	4.4867

The value of $x_3^{(2)}$ is found from the orthogonality relation to be -0.9473. We now have $\lambda_2 = 4.4867$, $x_1^{(2)} = 0.2170$, $x_2^{(2)} = 1$, $x_3^{(2)} = -0.9473$. Again eliminating x_3 and x_2 by the two orthogonality relations and setting $x_1 = 1$ we get directly $\lambda_3 = 2.1260$, $x_2^{(3)} = -0.5673$, $x_3^{(3)} = -0.3698$. Introducing the arbitrary constants we finally express the three sets of solutions as follows:

λ	x_1	x_2	x_3
8.3874	$0.8077c_1$	$0.7720c_1$	c_1
4.4867	$0.2170c_2$	c_2	$-0.9473c_2$
2.1260	c_3	$-0.5673c_3$	$-0.3698c_3$

It is of interest to note the first root obtained for λ was the largest root, the second was the second largest, and so on. This property is characteristic of the process.

Exercises

1. Solve

$$\begin{aligned} \lambda x_1 &= 2x_1 - x_2, \\ \lambda x_2 &= -x_1 + 2x_2 - x_3, \\ \lambda x_3 &= - x_2 + 2x_3. \end{aligned}$$

2. Solve

$$\begin{aligned} \lambda x_1 &= 4x_1 - 2x_2 + x_3, \\ \lambda x_2 &= -2x_1 + 4x_2 - 2x_3, \\ \lambda x_3 &= x_1 - 2x_2 + 5x_3. \end{aligned}$$

Chapter III
INTERPOLATION*

Interpolation may be regarded as a process for finding the value y of a function

$$y = f(x) \tag{1}$$

corresponding to any x, where the value of y is not to be computed directly from the function itself but is determined by means of certain values of the function which are already known. That is, given the pairs (x_1, y_1), (x_2, y_2), ..., (x_n, y_n) which satisfy (1), the problem is to find an approximate value of y corresponding to any selected value of x. The given values may constitute a table, such as a table of logarithms or natural sines; they may have been found indirectly, say from a series or from a differential equation; or they may have been obtained from experiments where the functional relationship is inferred but not mathematically expressed.

Strictly speaking the term "interpolation" is used only when the value of x in question does not fall outside the range of the given x's. Otherwise, the usual term is "extrapolation."

This chapter is devoted to an introductory study of methods of interpolation.

18. INTERPOLATING FUNCTIONS

An interpolating function may be defined as a function I(x) which contains in addition to the independent variable x a number of arbitrary constants or parameters in such a way that by suitable choice of the

*See Appendix C.

parameters the function $I(x)$ will assume assigned values for given values of the variable x. That is, if (x_1, y_1), (x_2, y_2), ..., (x_n, y_n) are n pairs of values, with the x's all distinct, $I(x_1) = y_1$, $I(x_2) = y_2$, ..., $I(x_n) = y_n$. Examples of interpolating functions of three different types are shown below for the case $n = 3$.

1) Polynomial type. Let $I(x)$ be defined by the equation

$$I(x) = y_1 \frac{(x - x_2)(x - x_3)}{(x_1 - x_2)(x_1 - x_3)} + y_2 \frac{(x - x_1)(x - x_3)}{(x_2 - x_1)(x_2 - x_3)} + y_3 \frac{(x - x_1)(x - x_2)}{(x_3 - x_1)(x_3 - x_2)}.$$

Considered as a function of x it is clear that $I(x)$ is a polynomial of second degree which could be reduced to the form

$$I(x) = A + Bx + Cx^2$$

in which the three coefficients A, B, C depend on the given quantities (x_1, y_1), (x_2, y_2), (x_3, y_3). By direct substitution we verify that

$$I(x_1) = y_1, \qquad I(x_2) = y_2, \qquad I(x_3) = y_3.$$

2) Rational type. Let

$$I(x) = \frac{\begin{vmatrix} y_1 & x_1 - x & y_1(x_1 - x) \\ y_2 & x_2 - x & y_2(x_2 - x) \\ y_3 & x_3 - x & y_3(x_3 - x) \end{vmatrix}}{\begin{vmatrix} 1 & x_1 - x & y_1(x_1 - x) \\ 1 & x_2 - x & y_2(x_2 - x) \\ 1 & x_3 - x & y_3(x_3 - x) \end{vmatrix}}$$

This may be reduced to a rational fraction

$$I(x) = \frac{A + Bx}{C + Dx}$$

where the constants A, B, C, D depend on (x_1, y_1), (x_2, y_2), (x_3, y_3). Note that actually only three independent constants occur since any one of the four which is not zero may be divided out of numerator and denominator. Again we see that

$$I(x_1) = y_1, \qquad I(x_2) = y_2, \qquad I(x_3) = y_3.$$

3) <u>Trigonometric type</u>. Let

$$I(x) = y_1 \frac{\sin\frac{1}{2}(x - x_2) \sin\frac{1}{2}(x - x_3)}{\sin\frac{1}{2}(x_1 - x_2) \sin\frac{1}{2}(x_1 - x_3)}$$

$$+ y_2 \frac{\sin\frac{1}{2}(x - x_1) \sin\frac{1}{2}(x - x_3)}{\sin\frac{1}{2}(x_2 - x_1) \sin\frac{1}{2}(x_2 - x_3)}$$

$$+ y_3 \frac{\sin\frac{1}{2}(x - x_1) \sin\frac{1}{2}(x - x_2)}{\sin\frac{1}{2}(x_3 - x_1) \sin\frac{1}{2}(x_3 - x_2)}.$$

Since

$$\sin\frac{1}{2}(x - x_2) \sin\frac{1}{2}(x - x_3) = \frac{1}{2}\cos\left(\frac{x_2 - x_3}{2}\right)$$

$$- \frac{1}{2}\cos\left(x - \frac{x_2 + x_3}{2}\right) = \frac{1}{2}\cos\left(\frac{x_2 - x_3}{2}\right)$$

$$- \frac{1}{2}\cos\left(\frac{x_2 + x_3}{2}\right) \cos x - \frac{1}{2}\sin\left(\frac{x_2 + x_3}{2}\right) \sin x$$

and similar relations hold for the other sine products it is apparent that I(x) may be expressed in the form

$$I(x) = A + B \cos x + C \sin x.$$

It is also evident that

$$I(x_1) = y_1, \qquad I(x_2) = y_2, \qquad I(x_3) = y_3.$$

The foregoing illustrations represent three types of interpolating functions; polynomial, rational, trigonometric. Obviously, countless other types might be devised. Hence, the practical question arises. What type is best?

Experience has shown that in general the polynomial interpolating function is the most useful, and for many reasons. Polynomials are simple in form, can be calculated by elementary operations, are free from singular points, are unrestricted as to range of values, may be differentiated or integrated without difficulty, and the coefficients to be determined enter linearly. For special cases, however, other types may be advantageous. If the given function is periodic, it is natural to employ trigonometric interpolation. If the given function becomes infinite in or near the range of given values then interpolation by means of rational functions is indicated.

This chapter is devoted to polynomial interpolation. It is convenient to cite here some important properties of polynomials.

I. A polynomial of degree n

$$y = c_0 + c_1 x + c_2 x^2 + \dots + c_n x^n \qquad (c_n \neq 0)$$

vanishes for n and only n values of x, real or complex, multiple values, if any, being counted according to their order of multiplicity.

II. If a polynomial of degree not exceeding n vanishes for more than n distinct values of x, the polynomial is identically zero.

III. If two polynomials of degree not greater than n are equal for more than n distinct values of x, the polynomials are identically equal.

IV. A polynomial of actual degree n may be factored in essentially only one way into the product of n linear factors $c_n (x - x_1)(x - x_2) \dots (x - x_n)$.

V. If $(x_o, y_o), (x_1, y_1), \dots, (x_n, y_n)$ are n + 1 pairs of values of x and y such that $x_o, x_1, \dots, x_n$ are all distinct there exists a unique polynomial $I_n(x)$ of degree not exceeding n such that $I_n(x_o) = y_o, I_n(x_1) = y_1, \dots, I_n(x_n) = y_n$.

The proposition V is so important that we give an outline of its proof. Let $L_j^{(n)}(x)$ denote the polynomial of degree n defined by the equation

$$L_j^{(n)}(x) = \frac{(x - x_o) \dots (x - x_{j-1})(x - x_{j+1}) \dots (x - x_n)}{(x_j - x_o) \dots (x_j - x_{j-1})(x_j - x_{j+1}) \dots (x_j - x_n)}.$$

The numerator contains all factors of the form $(x - x_i)$ <u>except</u> the factor $(x - x_j)$. The denominator is the value of the numerator with x replaced by x_j. Clearly the denominator does not vanish, since the x_i's are all distinct. Now let

$$I_n(x) = y_o L_o^{(n)}(x) + y_1 L_1^{(n)}(x) + \dots + y_n L_n^{(n)}(x).$$

Evidently $I_n(x)$ is a polynomial of degree not greater than n. If x be replaced by x_o, $L_o^{(n)}(x_o) = 1$, $L_1^{(n)}(x_o) = 0, \dots, L_n^{(n)}(x_o) = 0$, so that $I_n(x_o) = y_o$. Similarly, $I_n(x_1) = y_1$, etc. That $I_n(x)$ is unique follows directly from proposition III above.

19. LINEAR INTERPOLATION

The familiar process of linear interpolation which everyone learns in trigonometry is a special case of polynomial interpolation using two points and a polynomial of degree 1. It will be convenient to adopt the notation $I_{0, 1, \ldots, n}(x)$ to represent the interpolating polynomial of degree n determined for the n + 1 points $(x_0, y_0), (x_1, y_1), \ldots (x_n, y_n)$. Then in particular $I_{0, 1}(x)$ denotes the first degree polynomial for the points (x_0, y_0), (x_1, y_1), $I_{1, 2}(x)$ the first degree polynomial for (x_1, y_1), (x_2, y_2), $I_{1, 3, 5}(x)$ the second degree polynomial for (x_1, y_1), (x_3, y_3), (x_5, y_5), etc. Since in each case the polynomial is unique it is clear that the order of subscripts is immaterial, so that $I_{1, 2}(x) = I_{2, 1}(x)$, $I_{1, 3, 5}(x) = I_{3, 5, 1}(x)$, etc.

Of the various ways in which the linear interpolating polynomial may be expressed, the most convenient for machine calculation is probably the following:

$$I_{0, 1}(x) = \frac{1}{x_1 - x_0} \begin{vmatrix} y_0 & x_0 - x \\ y_1 & x_1 - x \end{vmatrix}$$

For the two-rowed determinant is easily evaluated on the machine by cross multiplication, and the result, while still on the dials, is divided by $x_1 - x_0$.

Example. Given $\sin \theta = 0.56432$ find $\cos \theta$.

From the trigonometric tables we have

i	$x_i = \sin \theta$	$y_i = \cos \theta$	$x_i - x$
0	0.56425	0.82561	-7
1	0.56449	0.82544	17

Evaluating the determinant and dividing by 24 we obtain $\cos \theta = 0.82556$. Note that any common factor may be removed from $(x_0 - x)$, $(x_1 - x)$, and $(x_1 - x_0)$. Hence, we may ignore the decimal point and treat those quantities as integers. Note also that common digits on the left in y_0 and y_1 will also occur in y. Thus in the example above we could have ignored the digits 825 and written only

$$\frac{1}{24}\begin{vmatrix} 61 & -7 \\ 44 & 17 \end{vmatrix} = 56$$

so that again $\cos \theta = 0.82556$. In the case of repeated interpolations with numbers carried to many figures this may effect a worth-while saving of labor.

Exercises

The student should apply the above method to interpolation in tables of logarithms, trigonometric functions, etc., until the process is thoroughly mastered.

20. AITKEN'S PROCESS OF ITERATION

When linear interpolation fails to give adequate accuracy more points may be used, with an interpolating polynomial of higher degree than the first. Consider the expression

$$y = \frac{\begin{vmatrix} I_{0,\,1}(x) & x_0 - x \\ I_{1,\,2}(x) & x_2 - x \end{vmatrix}}{x_2 - x_0},$$

Note first that this expression is formed for $I_{0,\,1}(x)$ and $I_{1,\,2}(x)$ in exactly the same manner as $I_{0,\,2}$ is formed for y_0 and y_2. Secondly, it is evidently a polynomial in x of degree 2. Finally, if $x = x_0$, $y = I_{0,\,1}(x_0) = y_0$; if $x = x_1$,

$$y = \frac{I_{0,\ 1}(x_1)(x_2 - x_1) - I_{1,\ 2}(x_1)(x_0 - x_1)}{x_2 - x_0}$$

$$= y_1\left[\frac{(x_2 - x_1) - (x_0 - x_1)}{x_2 - x_0}\right] = y_1;$$

and if $x = x_2$, $y = I_{1,\ 2}(x_2) = y_2$. From these results it follows that

$$y = I_{0,\ 1,\ 2}(x).$$

In other words, we may obtain the second degree interpolating polynomial $I_{0,\ 1,\ 2}(x)$ by applying linear interpolation to $I_{0,\ 1}(x)$ and $I_{1,\ 2}(x)$. Evidently we could obtain the same result from either of the two forms

$$I_{0,\ 1,\ 2}(x) = \frac{\begin{vmatrix} I_{0,\ 1}(x) & (x_1 - x) \\ I_{0,\ 2}(x) & (x_2 - x) \end{vmatrix}}{x_2 - x_1}$$

$$I_{0,\ 1,\ 2}(x) = \frac{\begin{vmatrix} I_{0,\ 2}(x) & (x_0 - x) \\ I_{1,\ 2}(x) & (x_1 - x) \end{vmatrix}}{x_1 - x_0}.$$

This process may be extended indefinitely. In general the interpolating polynomial for $n + 1$ points is obtained by linear interpolation applied to two different interpolating polynomials each formed for n of the given points. For example

$$I_{0,\ 1,\ 2,\ 3,\ 4,\ 5}(x) = \frac{\begin{vmatrix} I_{0,\ 1,\ 2,\ 3,\ 4}(x) & (x_4 - x) \\ I_{0,\ 1,\ 2,\ 3,\ 5}(x) & (x_5 - x) \end{vmatrix}}{x_5 - x_4}$$

$$= \frac{\begin{vmatrix} I_{0,\,1,\,2,\,3,\,5}(x) & (x_2 - x) \\ I_{0,\,1,\,3,\,4,\,5}(x) & (x_4 - x) \end{vmatrix}}{x_4 - x_2}$$

etc.

In Aitken's process an array of linear interpolations is set up according to the following scheme:

$$\begin{array}{llllll} x_0 & y_0 & & & & x_0 - x \\ x_1 & y_1 & I_{0,\,1}(x) & & & x_1 - x \\ x_2 & y_2 & I_{0,\,2}(x) & I_{0,\,1,\,2}(x) & & x_2 - x \\ x_3 & y_3 & I_{0,\,3}(x) & I_{0,\,1,\,3}(x) & I_{0,\,1,\,2,\,3}(x) & x_3 - x \end{array}$$

The calculations are easily performed on the calculating machine by cross multiplication and division, since the elements of the numerator determinant occur in the array in the same relative positions as in the determinant.

Example 1. From the tabulated values of the function Si(x) at intervals of 0.1 calculate Si(0.462).

The complete computation appears below. The columns headed x_i and y_i are taken from the table and the remaining values are calculated in accordance with the scheme above.

x_i	y_i	1st degree	2nd degree	3rd degree	$x_i - x$
0.3	0.29850				-162
0.4	0.39646	0.457195			-62
0.5	0.49311	0.456134	0.456537		38
0.6	0.58813	0.454900	0.456484	0.456557	138
0.7	0.68122	0.453502	0.456432	0.456557	238

The computations have been carried to one additional decimal place in order to reduce the accumulation of errors due to neglected digits. The line for $x = 0.7$ is not really needed but serves as a control, since the fact that the two entries in the 3rd degree column are identical is a partial check on the numerical computation and also indicates that a higher degree polynomial is unnecessary. Since the original data were given to five places, the interpolated value should also be rounded off to five places, giving Si(0.462) = 0.45656. The true value of Si(0.462) to seven places is 0.4565566, so that our interpolated value is really quite close to the true value.

<u>Example 2</u>. From a table of sin x for x given in radians at intervals of 0.1 find the value of sin 4.238.

The calculation is shown below

x	sin x	1st degree	2nd degree	3rd degree	4th degree	5th degree	$x_i - x$
4.0	-0.75680250						-238
4.1	.81827711	-.90311207					-138
4.2	.87157577	-.89338269	-.88968553				-38
4.3	.91616594	-.88323083	939401	-0.88957475			62
4.4	.95160207	-.87270824	912631	7928	-0.88957194		162
4.5	.97753012	-.86186885	888316	8390	1	-0.88957199	262

The value should be -0.88957200.

Exercises

1. From the table of values given in Example 1 calculate Si(0.45); Si(0.47); Si(0.43).
2. From the table of values given in Example 2 calculate sin 4.25; sin 4.275; sin 4.23.

21. NEVILLE'S METHOD OF ITERATION

A variation of the foregoing process is due to Neville. Its basic principle is the same as in Aitken's method but the computational setup is modified as follows:

x_0	y_0				$x_0 - x$
x_1	y_1	$I_{0,1}(x)$			$x_1 - x$
x_2	y_2	$I_{1,2}(x)$	$I_{0,1,2}(x)$		$x_2 - x$
x_3	y_3	$I_{2,3}(x)$	$I_{1,2,3}(x)$	$I_{0,1,2,3}(x)$	$x_3 - x$

Care must be exercised to select the correct entries from the column $x_i - x$ since these do not always lie in the same line of computation as the entries for the left-hand column of the determinant. For instance

$$I_{0,1,2,3}(x) = \frac{1}{x_3 - x_0}\begin{vmatrix} I_{0,1,2}(x) & x_0 - x \\ I_{1,2,3}(x) & x_3 - x \end{vmatrix}$$

Example. The problem given in Example 2 of Art. 20 is here worked by Neville's Method.

x	sin x	1st degree	2nd degree	3^{rd}degree	4th degree	5th degree	$x_i - x$
4.0	-0.75680250						-238
4.1	.81827711	-0.90311207					-138
4.2	.87157577	-0.89182926	-0.88968553				-38
4.3	.91616594	-0.88852003	54589	-0.88957475			62
4.4	.95160207	-0.89419554	59838	003	-0.88957194		162
4.5	.97753012	-0.90959863	42058	586	204	-0.88957199	262

Exercises

Solve the Exercises of Art. 20 by Neville's Method.

22. INVERSE INTERPOLATION

Inverse interpolation in a given table of values may usually be performed by Aitken's or Neville's Methods merely by the interchange of the roles of the dependent and independent variables in the computational scheme.

Example 1. Given a table of sin x for x in radians at intervals of 0.001 find x if $y = \sin x = 0.664$.

The computation for Aitken's method appears as follows:

y = sin x	x	1st degree	2nd degree	$y_i - y$
0.66313544	0.725			-86456
388361	6	0.72615557		-11639
463111	7	608	0.72615565	63111
537795	8	660	5	137795

The result is $x = 0.72615565$. It is apparent that the line for $x = 0.728$ was not needed, but serves as a control.

This is the usual method of inverse interpolation and is to be preferred whenever it is applicable. Since in effect we are representing the inverse function $x = f^{-1}(y)$ by means of an interpolating polynomial $I(y)$, this method of inverse interpolation is applicable only if $f^{-1}(y)$ is capable of being represented with sufficient accuracy by $I(y)$.

If it happens that the inverse interpolation is to be performed near a point where the tangent to the curve $y = f(x)$ is horizontal, the foregoing method is inapplicable because the inverse function $x = f^{-1}(y)$ has a vertical tangent and is not capable of being accurately represented by any polynomial. It is better therefore to obtain the direct interpolating polynomial $I(x)$ and then to solve the algebraic equation

$$I(x) = y$$

for the unknown x, the value of y being known. The methods of Articles 11 and 12 are easily adapted to the present situation, but with a slight modification due to the fact that in actual practice we do not obtain the explicit expression for $I(x)$.

If the value of y for which the corresponding x is desired lies between two consecutive given values y_0 and y_1 we may proceed as follows:

(a) Find a trial value x' of x by linear interpolation in the form

(1) $$x' = \frac{1}{y_1 - y_0}\begin{vmatrix} x_0 & y_0 - y \\ x_1 & y_1 - y \end{vmatrix}.$$

(b) Interpolate (by any method) to obtain the y' corresponding to x'.

(c) Calculate a second trial value x" by the formula

(2) $$x'' = \frac{1}{y' - y_0}\begin{vmatrix} x_0 & y_0 - y \\ x' & y' - y \end{vmatrix}.$$

(If y is nearer y_1 than it is to y_0, then x_1, y_1 should replace x_0, y_0 in (2)).

(d) Interpolate to obtain the y" corresponding to x".

(e) Calculate a third trial value x'" by the formula

(3) $$x''' = \frac{1}{y'' - y'}\begin{vmatrix} x' & y' - y \\ x'' & y'' - y \end{vmatrix}.$$

The process is continued until an x is reached which by interpolation gives the required value of y.

Example 2. From the following table of y = sin x in radians find the value of x corresponding to y = 0.96968.

x	y
1.2	0.93204
1.3	0.96356
1.4	0.98545
1.5	0.99749
1.6	0.99957

Following the steps proposed above we have

(a) x' = 1.328.

(b) y' = 0.97067 by Aitken's method.

(c) x" = 1.324.

(d) y" = 0.96970.

(e) x'" = 1.3239, which is taken as correct.

For inverse interpolation very near a maximum or a

minimum of f(x) it is usually advantageous to modify the first step (a) in the method just outlined by using a quadratic rather than a linear formula to secure the first trial value (or values). Let (x_0,y_0), (x_1,y_1), (x_2,y_2) be three consecutive given points nearest to the point where the interpolation is to be performed. Then an appropriate quadratic formula is

$$(4)\qquad A(x - x_1)^2 + B(x - x_1) + C(y - y_1) = 0,$$

in which

$$A = (x_2 - x_1)y_0 + (x_0 - x_2)y_1 + (x_1 - x_0)y_2,$$

$$B = -(x_2 - x_1)^2 y_0 + (x_2 - 2x_1 + x_0)(x_2 - x_0)y_1 + (x_1 - x_0)^2 y_2,$$

$$C = (x_2 - x_1)(x_2 - x_0)(x_1 - x_0).$$

It may be verified by direct substitution that (4) is satisfied by the three pairs of values (x_0,y_0), (x_1,y_1) and (x_2,y_2).

In the important case where the values of x are equally spaced with interval h equation (4) assumes the simple form

$$(5)\qquad (y_2 - 2y_1 + y_0)s^2 + (y_2 - y_0)s + 2(y_1 - y) = 0,$$

where $s = (x - x_1)/h$

Example 3. From the following tabulated values of the function $y = \mathrm{Si}(x)$ determine x corresponding to $y = 1.8519336917$.

x	y
3.12	1.8518625083
3.13	9156107
3.14	9366481
3.15	9258225
3.16	8833367

Using the values for x = 3.13, 3.14, 3.15, we set up equation (5), which turns out to be

$$318630\, s^2 - 102118\, s - 59128 = 0.$$

The two roots, taken to two decimal places, are $s = -0.30$ and $s = 0.62$, to which correspond $x' = 3.1370$ and $x' = 3.1462$. We now treat each one of these separately by the methods used in Example 2 in order to obtain closer approximations.

Direct interpolation with $x' = 3.1370$ gives the desired value of y immediately and no further refinement is required.

For the second value we obtain the successive approximations

$$x' = 3.1462, \quad y' = 1.8519336768,$$
$$x'' = 3.14619, \quad y'' = 1.8519336915,$$
$$x''' = 3.1461899 \quad y''' = 1.8519336917.$$

Note that the curve is so flat in the vicinity of the desired points that x cannot be determined accurately beyond seven decimal places. Accordingly we take $x = 3.1370000$ and $x = 3.1461899$ as the answers to the problem.

If the roots of the quadratic equation are imaginary we may infer that either the given problem has no solution or has two or more solutions so close together that our quadratic does not fit closely enough to locate them.

It is important to observe that this process of inverse interpolation by successive approximations is in no way dependent on the particular method used for performing the direct interpolations. Thus it may be used equally well with Aitken's, Neville's, or Lagrange's methods, or with any of the various methods of later chapters based on differences.

Exercises

1. Using the values given in Example 1 of Article 20, find x if Si(x) - 0.45.

2. Using the values in Example 2, Article 20, find x if $\sin x = 0.85000000$. If $\sin x = 0.88432716$.

3. Carry out in full the steps outlined in Example 2.

4. Carry out the steps outlined in Example 3.

5. With the data of Example 2 find x for $y = 0.98000$.

6. With the data of Example 3 find x for $y = 1.8519200000$.

7. From the values of $y = \sin x$ given below

x	y
1.569	0.99999839
1.570	968
1.571	998
1.572	928
1.573	757
1.574	487

a) find x for $y = 0.99999990$;

b) find x for $y = 1.00000000$.

How many places in the value found for x are reliable?

23. THE ERROR IN POLYNOMIAL INTERPOLATION

The use of the interpolating polynomial to obtain a value of the function for an x different from the given x's for which the polynomial was constructed is based on the assumption that in the interval under consideration the polynomial is a good approximation to the function. It is essential that we have a criterion for checking this assumption. Such a criterion is given by the following:-

Theorem. *If $f(x)$ has continuous derivatives up to order $n + 1$ in an interval and if $P_n(x)$ is a polynomial of degree n such that $P_n(x) = f(x)$ for $x = x_0, x_1, \ldots, x_n$ in this interval then*

$$f(x) - P_n(x) = \frac{f^{(n+1)}(s)(x - x_0)(x - x_1) \ldots (x - x_n)}{(n + 1)!}$$

in which s lies between the greatest and least of the numbers $x, x_0, x_1, \ldots x_n$.

Proof. Let the auxiliary function $F(x)$ be defined by the equation

(1) $F(x) = f(x) - P_n(x) - R(x - x_0)(x - x_1) \ldots (x - x_n)$,

in which R is a constant so chosen that

(2) $f(x') - P_n(x') - R(x' - x_0)(x' - x_1) \ldots (x' - x_n) = 0$

where x' is some arbitrarily chosen value of x. Clearly the function $F(x)$ vanishes at n+2 distinct points, $x_0, x_1, \ldots, x_n, x'$. By Rolle's Theorem $F'(x)$ vanishes at least at n+1 distinct points, $F''(x)$ at n points, and so on, and in particular $F^{(n+1)}(x)$ vanishes at least at one point s lying between the greatest and least of the numbers $x_0, x_1, \ldots, x_n, x'$. We differentiate (1) n+1 times and obtain

(3) $$F^{(n+1)}(x) = f^{(n+1)}(x) - (n+1)!R,$$

because the $(n+1)^{th}$ derivative of $P_n(x)$ is zero, since the degree of $P(x)$ does not exceed n, while the product $(x - x_0)(x - x_1) \ldots (x - x_n)$, when multiplied out, starts with the term x^{n+1} followed by terms of lower degree. Now into (3) we substitute $x = s$, the value for which $F^{(n+1)}(x)$ vanishes, and solve for R, obtaining

$$R = f^{(n+1)}(s)/(n+1)!$$

Putting this value of R into (2) and transposing we have (neglecting the primes, which are no longer needed)

(4) $$f(x) - P_n(x) = \frac{f^{(n+1)}(s)(x - x_0)(x - x_1) \ldots (x - x_n)}{(n+1)!}.$$

This is the desired formula for the error. Its use requires a knowledge of the magnitude of $f^{(n+1)}(x)$ in the interval under consideration. When $f(x)$ is a definite mathematical function which can be differentiated this requirement causes no particular trouble, but in the case of experimentally determined functions some assumption regarding the size of $f^{(n+1)}(x)$ must be made. (See Article 36.)

Example 1. Let us find the error committed in approximating to sin x by a polynomial of fifth degree which conincides with sin x at the points 0°, 5°, 10°, 15°, 20°, 25°.

Here $f(x) = \sin x$, and $f^{(6)}(x) = -\sin x$, so $f^{(6)}(x)$ does not exceed 1 in numerical value. Then

$$\left|\sin x - P(x)\right| \leq \frac{x(x - \frac{\pi}{36})(x - \frac{\pi}{18})(x - \frac{\pi}{12})(x - \frac{\pi}{9})(x - \frac{5\pi}{36})}{6!}$$

since $5^\circ = \frac{\pi}{36}$ radians. Setting $x = 12^\circ 30'$ (= 0.21816 rad. for example, and evaluating we find

$$|\sin x - P(x)| < 0.00000000216.$$

Again setting $x = 2^\circ$, we have

$$|\sin x - P(x)| < 0.000000010.$$

Example 2. Find the error in approximating to $f(x) = \log_{10} x$ by a polynomial of the 5th degree which coincides with $\log_{10} x$ at $x = 1, 2, 3, 4, 5, 6$.

Here we find upon differentiating six times that

$$f^{(6)}(x) = -\frac{\log_{10} e \cdot 5!}{x^6}.$$

Then

$$|\log_{10} x - P(x)|$$
$$\leq \frac{\log_{10} e \cdot (x - 1)(x - 2)(x - 3)(x - 4)(x - 5)(x - 6)}{6s^6}$$

If x is in the interval $1 \leqq x \leqq 6$, then s also is in this interval, and in the worst case possible, namely, when $s = 1$, we have

$$|\log x - P(x)|$$
$$< (.0724)(x - 1)(x - 2)(x - 3)(x - 4)(x - 5)(x - 6).$$

In particular if $x = 1.5$ we have

$$|\log_{10} x - P(x)| < 1.07.$$

Actually the error is considerably less than 1, but how much less cannot be determined from the error formula.

Example 3. As a third example let us take the function $f(x) = \log_{10} x$ of Example 2, but approximate it by using the points $x = 10, 11, 12, 13, 14, 15$. Here we find that for $x = 10.5$ the error is less than 0.00001. The reason for the great improvement in accuracy is that in this example the least value of s is 10, while in the former the least value of s was 1. In other words the 6th derivative is roughly a million times larger in the second example than in the third.

Example 4. To show the effect of a shorter interval upon the error, consider again the function $f(x) = \log_{10} x$ and let us approximate by a polynomial of fifth degree coinciding with $\log_{10} x$ at 1, 1.01, 1.02, 1.03, 1.04, 1.05.

Here we have

$$|\log_{10} x - P(x)|$$
$$< (.0724(x-1)(x-1.01)(x-1.02)(x-1.03)(x-1.04)(x-1.05).$$

At $x = 1.015$ for example, we find

$$|\log_{10} x - P(x)| < (3.6)10^{-13}.$$

Thus we see that P(x) will represent $\log_{10} x$ accurately to about 12 places of decimals for x between 1 and 1.05.

Exercises

1. A table of natural sines is given with entries for every degree.
   What is the maximum error of linear interpolation (i.e. approximation by a polynomial of degree 1)?
2. In Exercise 1, what is the error if entries are given for every minute?
3. In Exercise 1, what is the error if entries are given for every second?
4. In a table of natural logarithms of numbers from 1 to 10 entries are given at intervals of 0.001. What is the maximum error of linear interpolation?
5. Show that interpolation is inaccurate in a table of log sin x in the immediate vicinity of x = 0.
6. Show that interpolation is inaccurate in a table of natural tangents in the vicinity of $x = 90^{\circ}$.
7. A table of the probability integral

$$y = \frac{2}{\sqrt{\pi}} \int_0^x e^{-x^2} dx$$

   from x = 0 to x = 3 has entries at intervals of 0.001. What is the maximum error of linear interpolation.
8. A table of e^x from 0 to 1 has entries at intervals of 0.01. What is the maximum error of linear interpolation?
9. Find by formula a parabola through the points (0,0), $(\frac{\pi}{2}, 1)$, $(\pi, 0)$. (These are three values of y = sin x.) Calculate the error theoretically. Then compare with the actual error.
10. Find the error in approximating tan x with a polynomial of third degree coinciding at $x = 0^{\circ}, 1^{\circ}, 2^{\circ}, 3^{\circ}$. Find the error when $x = 30'$; $x = 1^{\circ}30'$; $x = 2^{\circ}30'$.

24. LAGRANGE'S INTERPOLATION FORMULA

It has already been shown in Art. 18 that the formula

$$(1) \quad I(x) = y_0 L_0^{(n)}(x) + y_1 L_1^{(n)}(x) + \ldots + y_n L_n^{(n)}(x),$$

where

$$(2) \quad L_j^{(n)}(x) = \frac{(x - x_0)\ldots(x - x_{j-1})(x - x_{j+1})\ldots(x - x_n)}{(x_j - x_0)\ldots(x_j - x_{j-1})(x_j - x_{j+1})\ldots(x_j - x_n)},$$

gives the interpolating polynomial for the $n + 1$ points (x_0, y_0), (x_1, y_1), . . ., (x_n, y_n). Formula (1) is known as Lagrange's interpolation formula and the coefficients (2) are the Lagrangian coefficients.

These coefficients have several properties which merit attention. First of all, the Lagrangian coefficient $L_j^{(n)}(x)$ formed for the $n + 1$ points (x_0, y_0), (x_1, y_1), . . ., (x_n, y_n) is a polynomial of degree n which vanishes at $x = x_0$, . . ., $x = x_{j-1}$, $x = x_{j+1}$, . . ., $x = x_n$, but at $x = x_j$ it assumes the value 1. Second, the form of $L_j^{(n)}(x)$ as given by (2) shows that the Lagrangian coefficients depend only on the given x's and are entirely independent of the y's. Third, the Lagrangian coefficients possess the useful property that their <u>form</u> remains invariant if the variable x is replaced by a new variable s through the transformation $x = hs + a$, where h and a are constants, $h \neq 0$. For by making the substitutions $x = hs + a$, $x_0 = hs_0 + a$, $x_1 = hs_1 + a$ we may verify directly that

$$L_0^{(n)}(x) = \frac{(x - x_1)(x - x_2)\ldots(x - x_n)}{(x_0 - x_1)(x_0 - x_2)\ldots(x_0 - x_n)}$$

$$= \frac{(s - s_1)(s - s_2)\ldots(s - s_n)}{(s_0 - s_1)(s_0 - s_2)\ldots(s_0 - s_n)}$$

and similarly for the other coefficients. Fourth, if $\Pi_{n+1}(x)$ denotes the product of the $n + 1$ factors $(x - x_0)$, $(x - x_1)$, . . ., $(x - x_n)$

$$\Pi_{n+1}(x) = (x - x_0)(x - x_1) \dots (x - x_n)$$

we obtain, by differentiating the product and then setting $x = x_0$,

$$\Pi'_{n+1}(x_0) = (x_0 - x_1)(x_0 - x_2) \dots (x_0 - x_n).$$

Similarly

$$\Pi'_{n+1}(x_1) = (x_1 - x_0)(x_1 - x_2) \dots (x_1 - x_n), \text{ etc.}$$

Hence, it is clear that

$$L_j^{(n)}(x) = \frac{\Pi_{n+1}(x)}{(x - x_j)\Pi'_{n+1}(x_j)} \tag{3}$$

Because of the invariant property referred to above, if $x = hs + a$, equation (3) may be expressed as

$$L_j^{(n)}(x) = \frac{\Pi_{n+1}(s)}{(s - s_j)\Pi'_{n+1}(s_j)} \tag{3'}$$

where in this case

$$\Pi_{n+1}(s) = (s - s_0)(s - s_1) \dots (s - s_n).$$

Equation (3) makes possible a convenient and compact form for writing Lagrange's formula, as follows:

$$I(x) = \Pi_{n+1}(x)\sum_{j=0}^{n} \frac{y_j}{(x - x_j)\Pi'_{n+1}(x_j)} \tag{4}$$

It will be noted that Lagrange's formula contains the values y_0, y_1, . . ., y_n explicitly, a fact of considerable importance for subsequent applications. In this respect, it possesses certain advantages over Aitken's or Neville's processes, where the original values of the y's are lost sight of in the course of the computation.

On the other hand, Lagrange's formula has the disadvantage that in itself it gives no indication of the number of points required to secure a desired degree of accuracy, whereas Aitken's and Neville's processes, by convergence toward a fixed value, do give such an indication.

When we wish to express the interpolating polynomial explicitly as a function of the variable x rather than to compute a value of the polynomial for a given numerical value of x the use of Lagrange's formula is advantageous.

Example 1. Construct a polynomial having the values given by the table

x :	1	3	4	6
y :	-7	5	8	14

By substituting the given values of x and y into formula (1) we have

$$y = (-7)\frac{(x-3)(x-4)(x-6)}{(-2)(-3)(-5)} + (5)\frac{(x-1)(x-4)(x-6)}{(2)(-1)(-3)}$$

$$+ (8)\frac{(x-1)(x-3)(x-6)}{(3)(1)(-2)} + (14)\frac{(x-1)(x-3)(x-4)}{(5)(3)(2)},$$

which upon a simplification reduces to

$$y = \frac{1}{5}(x^3 - 13x^2 + 69x - 92).$$

It is readily verified by substitution that this polynomial satisfies the required conditions.

If the interpolation formula is to be used simply for calculating values of the function corresponding to assigned values of x it is unnecessary to reduce the polynomial to the simplified form as the substitution may be made directly into the original expression in product form. For some purposes, however, especially if we wish to differentiate or integrate the interpolating polynomial, it is desirable to have the Lagrangian coefficients expressed in terms of powers of the variable x. This may be done in a systematic manner by use of formula (3).

<u>Example 2.</u> Obtain the Lagrangian coefficients for the points

$$x = 3.4, \quad 3.5, \quad 3.7, \quad 3.8, \quad 4.0.$$

It is convenient first of all to introduce a new variable s by the relation $x = \frac{1}{10}s + 3.4$ so that the values of s corresponding to the given x's will be the integers 0, 1, 3, 4, 6.

Then using (3') we set

$$\prod(s) = s(s - 1)(s - 3)(s - 4)(s - 6)$$

$$= s^5 - 14s^4 + 67s^3 - 126s^2 + 72s,$$

and compute terms of the form

$$\frac{\prod(s)}{s - s_j}$$

by synthetic division. Since $\prod'(s_j) = \left.\frac{\prod(s)}{s - s_j}\right|_{s=s_j}$ the value of $\prod'(s_j)$ is now obtainable by synthetic division followed by substitution. For example for $s = 4$ we arrange the computation in the form

4	1	- 14	+ 67	- 126	+ 72	- 0
		4	- 40	108	- 72	0
	1	- 10	27	- 18	0	0
		4	- 24	12	- 24	
	1	- 6	3	- 6	- 24	

The numbers in the third line are the coefficients in the

polynomial $\frac{\prod(s)}{s - 4}$, and the last number in the last line is $\prod'(4)$, so that

$$L_3^{(4)}(x) = -\frac{1}{24}(s^4 - 10s^3 + 27s^2 - 18s).$$

In this manner we get all five Lagrangian coefficients as follows:

$$L_0^{(4)}(x) = \frac{1}{72}(s^4 - 14s^3 + 67s^2 - 126s + 72),$$

$$L_1^{(4)}(x) = -\frac{1}{30}(s^4 - 13s^3 + 54s^2 - 72s),$$

$$L_2^{(4)}(x) = \frac{1}{18}(s^4 - 11s^3 + 34s^2 - 24s),$$

$$L_3^{(4)}(x) = -\frac{1}{24}(s^4 - 10s^3 + 27s^2 - 18s),$$

$$L_4^{(4)}(x) = \frac{1}{180}(s^4 - 8s^3 + 19s^2 - 12s),$$

where x and s are related by the equation $x = \frac{1}{10}s + 3.4$

When it is desired to compute the Lagrangian coefficients for some given numerical value of the variable x, it is advantageous to arrange the computation in a compact and systematic form. In the case where the values $x_0, x_1, \ldots, x_n$ are not equally spaced, the following scheme is suggested, and is illustrated for the case $n = 3$. We first set up the square array of differences

$$\begin{array}{cccc} \underline{x - x_0} & x_0 - x_1 & x_0 - x_2 & x_0 - x_3 \\ x_1 - x_0 & \underline{x - x_1} & x_1 - x_2 & x_1 - x_3 \\ x_2 - x_0 & x_2 - x_1 & \underline{x - x_2} & x_2 - x_3 \\ x_3 - x_0 & x_3 - x_1 & x_3 - x_2 & \underline{x - x_3}. \end{array}$$

Let the product of the numbers in the principal diagonal be denoted by $\prod(x)$; let the product of numbers in the the first row be D_0, second row D_1, etc. Then it is evident that

$$L_0^{(3)}(x) = \Pi(x)/D_0,\ L_1^{(3)}(x) = \Pi(x)/D_1,\ L_2^{(3)}(x) = \Pi(x)/D_2,\ L_3^{(3)}(x) = \Pi(x)/D_3,$$

and the interpolated value y is therefore

$$y = \Pi(x)S$$

where

$$S = y_0/D_0 + y_1/D_1 + y_2/D_2 + y_3/D_3.$$

The extension to the general case is obvious.

Example 3. The values of sin x for certain angles 0°, 30°, 45°, etc. can be calculated by elementary methods, giving us the table:

x	sin x	x	sin x	x	sin x
0°	0.000000	60°	0.866025	135°	0.707107
30°	0.500000	90°	1.000000	150°	0.500000
45°	0.707107	120°	0.866025	180°	0.000000

From this table obtain the value of sin 105° by interpolation.

Here it is convenient to make the change of variable x = 15s, so that the given values of s are 0, 2, 3, 4, 6, 8, 9, 10, 12, and s itself is 7. The complete computation is arranged as shown:

									D_i	y_i	y_i/D_i
7	-2	-3	-4	-6	-8	-9	-10	-12	8709120	0.000000	0
2	5	-1	-2	-4	-6	-7	-8	-10	-268800	0.500000	-18601.10^{-10}
3	1	4	-1	-3	-5	-6	-7	-9	68040	0.707107	103925 "
4	2	1	3	-2	-4	-5	-6	-8	-46080	0.866025	-187939 "
6	4	3	2	1	-2	-3	-4	-6	20736	1.000000	482253 "
8	6	5	4	2	-1	-1	-2	-4	15360	0.866025	563818 "
9	7	6	5	3	1	-2	-1	-3	-34020	0.707107	-207850 "
10	8	7	6	4	2	1	-3	-2	161280	0.500000	31002 "
12	10	9	8	6	4	3	2	-5	-6220800	0.000000	0

$\Pi(x) = 12600$ $\quad$ $S = 766608.10^{-10}$

$y = 0.965926$

This interpolated value agrees with the true value of $\sin 105^{\circ}$ to six places of decimals.

Exercises

1. Obtain the Lagrangian coefficients in polynomial form for the four points whose abscissas are $x = 2, 4, 6, 8$.
2. Determine the polynomial passing through the points $(2, 0), (4, 3), (6, 5), (8, 4)$.
3. From a table of common logarithms obtain log 125, log 126, log 128, log 129. Calculate log 127 by interpolation.

25. EQUALLY SPACED POINTS. TABLES OF LAGRANGIAN COEFFICIENTS

In the important case in which the values of the x's are equally spaced so that

$$x_1 - x_0 = x_2 - x_1 = \dots = x_n - x_{n-1} = h$$

the method of interpolating by means of Lagrange's formula as explained in Example 3 of Art. 24 can be greatly simplified. For in this case we may set $x = hs + x_0$ and find that $s_0 = 0$, $s_1 = 1$, ..., $s_n = n$. Then the product $\prod_{n+1}(s)$ becomes

$$\prod_{n+1}(s) = s(s - 1) \dots (s - n)$$

from which we may show that

$$\prod{}'_{n+1}(j) = (-1)^{n-j} j!(n - j)!$$

Using this fact, we now see that equations (4) of Art. 24 can be written

$$(1) \qquad I(s) = \frac{(-1)^n s(s - 1) \dots (s - n)}{n!} \sum_{j=0}^{n} \frac{(-1)^j \binom{n}{j} y_j}{s - j},$$

in which $\binom{n}{j}$ as usual denotes the binomial coefficient

$$\binom{n}{j} = \frac{n!}{j!\,(n-j)!}$$

The numerical procedure is illustrated by

Example 1. From a table of cos x with x in radians at intervals of 0.1, calculate cos x for $x = 5.347$.

We take eight entries from the table, four on each side of the required value, and let $x = 0.1s + 5$. The calculation is set up below and is self-explanatory in view of formula (1).

Required: cos x for $x = 5.347$.

x_i	$\cos x_i$	s_i	$s - s_i$	$(-1)^i\binom{n}{i}$	$(-1)^i\binom{n}{i}\frac{\cos x_i}{s - s_i}$
5.0	0.283662185	0	3.47	-1	-0.08174702
5.1	0.377977743	1	2.47	7	1.07119198
5.2	0.468516671	2	1.47	-21	-6.69309530
5.3	0.554374336	3	0.47	35	41.28319523
5.4	0.634692876	4	-0.53	-35	41.91368048
5.5	0.708669774	5	-1.53	21	-9.72684003
5.6	0.775565879	6	-2.53	-7	2.14583444
5.7	0.834712785	7	-3.53	1	-0.23646254

$n! = 5040$, $\Pi(s) = 42.8848749$, $S = 69.67575724$

$$\cos x = \frac{\Pi(s)S}{n!} = 0.592864312.$$

For the case of equally-spaced points, Tables of Lagrangian coefficients are available from which it is usually possible to find directly the numerical values of the coefficients, so that the actual labor of interpolation is reduced to multiplication of the y's by the coefficients, followed by a summation.

The most complete table now available is "Tables of

Lagrangian Interpolation Coefficients", prepared by the Mathematical Tables Project under the sponsorship of the National Bureau of Standards and published by Columbia University Press, New York, 1944. The use of this table will be indicated by applying it to Example 1 above.

Example 2. Find the value of cos 5.347 from the data of Example 1, using tables of coefficients.

For the case of eight given points, as in this example, four on each side of the desired value, the coefficients which we have designated by $L_j^{(7)}(x)$ are called A_{j-3} in the Tables so that the interpolation formula reads (using our notation for the y's)

$$I(x) = y_0A_{-3} + y_1A_{-2} + y_2A_{-1} + y_3A_0 + y_4A_1 + y_5A_2 + y_6A_3 + y_7A_4$$

Instead of the parameter s which we have used, the Tables employ parameter p, given by

$$x = hp + x_3$$

or in this case

$$x = 0.1p + 5.3$$

so that the desired value of p is 0.47. In Table VI for eight points the coefficients for p = 0.47 are found on pp. 352 - 353. Taking the given coefficients from the table we have:

x	y	A
5.0	0.283662185	-0.0024521336
5.1	0.377977743	0.0241143021
5.2	0.468516671	-0.1215557679
5.3	0.554374336	0.6336417689
5.4	0.634692876	0.5619087384
5.5	0.708669774	-0.1167888751
5.6	0.775565879	0.0235424215
5.7	0.834712785	-0.0024104543

Upon multiplication and summation, we get y = 0.592864312.

Exercises

Using the above values of cos x, calculate cos 5.356, cos 5.328, cos 5.361. (The Tables should be used if available.)

SUMMARY

This chapter deals with three methods of polynomial interpolation, Aitken's, Neville's and Lagrange's. Other methods will be treated in the chapters on differences. Since, as has been shown, the interpolating polynomial for a given set of points is unique, the question may well be asked, "Why not always use just one method to compute it?" The answer is somewhat complex, depending partly on the given data, partly on the equipment available (calculating machines, tables, etc.), and partly on the habits and preferences of the computer. For instance, a computer without a calculating machine would probably shy away from the lengthy multiplications and divisions involved in the methods of this chapter and resort instead to the method of differences which, as will appear later, depends more on additions and subtractions.

Table IV at the end of this book gives Lagrange's coefficients for five equally spaced points. This may serve to furnish practice in working exercises if a better table is not available.

Chapter IV
NUMERICAL DIFFERENTIATION AND INTEGRATION*

In Chapter III the interpolating polynomial was defined and methods were developed for obtaining the polynomial. We now turn to the problem of finding the derivative or the integral of a function $y = f(x)$ which may not be given in explicit mathematical form but for which we know pairs of values (x_0, y_0), (x_1, y_1), . . ., (x_n, y_n). An obvious procedure is first to obtain the interpolating polynomial and then to differentiate or integrate the latter, in the hope of thus securing a sufficiently good approximation to the derivative or integral of the given function. In this chapter we derive formulas for carrying out this procedure and for estimating the error.

26. NUMERICAL DIFFERENTIATION

Let I(x) denote the interpolating polynomial for the function f(x) at the points (x_0, y_0), (x_1, y_1), . . ., (x_n, y_n). Then by the theorem of Art. 23, we have

$$f(x) = I(x) + \frac{f^{(n+1)}(z)}{(n+1)!}(x - x_0)(x - x_1)\ .\ .\ .\ (x - x_n)$$

where we imply that the $(n + 1)$th derivative exists and is continuous over the range of values of x under consideration, and where z denotes some value of x in this range.

Differentiation of the above equation gives, formally

$$(1)\quad \frac{d}{dx} f(x) = \frac{d}{dx} I(x) + \frac{f^{(n+1)}(z)}{(n+1)!}\frac{d}{dx}\Big[(x - x_0)(x - x_1)\ .\ .\ .\ (x - x_n)\Big] + \left[\frac{d}{dx}\ \frac{f^{(n+1)}(z)}{(n+1)!}\right](x - x_0)(x - x_1)\ .\ .\ .\ (x - x_n).$$

*See Appendix C.

Since we have no way of knowing how z depends on x we are unable to evaluate the expression $\frac{d}{dx}f^{(n+1)}(z)$ even if we assume from now on that $f^{(n+2)}(x)$ exists. However, in the important special case where we want to calculate the derivative at $x = x_j$, (x_j being one of the given values $x_0, x_1, \ldots, x_n$) we may substitute $x = x_j$ in (1) and obtain

$$(2) \qquad \frac{d}{dx}f(x)_{x=x_j} = \frac{d}{dx} I(x)_{x=x_j} + \frac{f^{(n+1)}(z)}{(n+1)!}(x_j - x_0)(x_j - x_1)\ldots(x_j - x_n)$$

where the factor $(x_j - x_j)$ does not occur in the product on the right, and the troublesome term $\frac{d}{dx} f^{(n+1)}(z)$ has dropped out. Equation (2) provides an estimate of the error in approximating the derivative of a function by means of the derivative of the interpolating polynomial for the case where the value of the derivative is sought at one of the given points.

If the interpolating polynomial I(x) is expressed by means of Lagrange's formula in the form given by equation (4) of Art. 24, we obtain by differentiation

$$(3) \qquad \left[\frac{d}{dx} I(x)\right]_{x=x_j} = \Pi'_{n+1}(x_j)\sum_{k=0}^{n}(k\neq j)\frac{y_k}{(x_j - x_k)\Pi'_{n+1}(x_k)} + \sum_{k=0}^{n}(k\neq j)\frac{y_j}{x_j - x_k}$$

This formula may be adapted to numerical calculation by an obvious modification of the setup used in Example 3 of Art. 24.

Example 1. From the data given in Example 3, Art. 24, calculate $\frac{d}{dx}\sin x$ at $x = 45^{\circ}$, and estimate the error.

Here again we make the change $x = 15s$, and find the value of the derivative at $s = 3$, recalling that $\frac{d}{dx}\sin x = \frac{1}{15}\frac{d}{ds}\sin x$. The calculation is shown below. Note that according to formula (3) the coefficient of y_2 must be computed separately.

									$-(s_i - 3)\Pi'(s_i)$	y_i	$\frac{-y_i}{(s_i - 3)\Pi'(s_i)}$
	-2	-3*	-4	-6	-8	-9	-10	-12	3732480	0.000000	0
2		-1*	-2	-4	-6	-7	-8	-10	-53760	0.500000	-930060.10^{-11}
3	1		-1	-3	-5	-6	-7	-9		0.707107	
4	2	1*		-2	-4	-5	-6	-8	15360	0.866025	5638184.10^{-11}
6	4	3*	2		-2	-3	-4	-6	-62208	1.000000	-1607510.10^{-11}
8	6	5*	4	2		-1	-2	-4	76800	0.866025	1127637.10^{-11}
9	7	6*	5	3	1		-1	-3	-102060	0.707107	-692835.10^{-11}
10	8	7*	6	4	2	1		-2	376320	0.500000	132866.10^{-11}
12	10	9*	8	6	4	3	2		-11197440	0.000000	0

$\Pi'(3) = 17010$

$S = 36682.10^{-11}$

$\Pi'(3)S = 0.6239758$

Coefficient of $y_2 = \frac{1}{3} + \frac{1}{1} - \frac{1}{1} - \frac{1}{3} - \frac{1}{5} - \frac{1}{6} - \frac{1}{7} - \frac{1}{9} = -\frac{1173}{1890}$

$\frac{d}{ds}\sin x = \Pi'(3)S - \frac{1173}{1890}(0.707107) = 0.1851206,$

$\frac{d}{dx}\sin x = 0.01234137.$

The entry for $+(s_i - 3)\Pi'(s_i)$ is the product of the numbers in the same line to the left, the starred number (in the column where $s = 3$) being used twice. The quantity $\Pi'(3)$ is the product of the numbers in the third row. The coefficient of y_2 is the sum of reciprocals of numbers in the third row.

The error is calculated from the expression

$$E = \frac{f^{(9)}(z)}{(9)!}(x_2 - x_0)(x_2 - x_1)(x_2 - x_3)\ .\ .\ .\ (x_2 - x_8).$$

Since $f(x) = \sin x$, $f^{(9)}(x) = (\frac{\pi}{180})^9 \cos x$, and we have

$$E = \left(\frac{\pi}{180}\right)^9 \cos z \cdot 15^8(3)(1)(-1)(-3)(-5)(-6)(-7)(-9)$$

whence

$$|E| < 0.00000002 \text{ approximately.}$$

The error of our result above is actually greater than this, due probably to neglect of digits beyond the sixth in the values of y.

Exercises

1. In the example above find $\frac{d}{dx}\sin x$ for $x = 90^\circ$; for $x = 30^\circ$; for $x = 0^\circ$.
2. Estimate the derivative of f(x) at x = 4, given the values (1, 7), (3, 8), (4, 9), (6, 11).

27. DIFFERENTIATION FORMULAS FOR EQUALLY-SPACED POINTS

For the case of equally-spaced points with constant interval h, the coefficients in formula (3) of Art. 26 can be determined once for all for any given n and j. A partial list of such formulas is given below, together with the remainder term for each. In the expression for the remainder term the notation $y^{(n+1)}$ means the (n + 1)th derivative of y taken at some point x = z lying between x_0 and x_n.

List of Differentiation Formulas

n = 2. Three points

$$y_0' = \frac{1}{2h}(-3y_0 + 4y_1 - y_2) + \frac{h^2}{3}y^{(3)}$$

$$y_1' = \frac{1}{2h}(-y_0 + y_2) - \frac{h^2}{6}y^{(3)}$$

$$y_2' = \frac{1}{2h}(y_0 - 4y_1 + 3y_2) + \frac{h^2}{3}y^{(3)}$$

n = 3. Four points

$$y_0' = \frac{1}{6h}(-11y_0 + 18y_1 - 9y_2 + 2y_3) - \frac{h^3}{4}y^{(4)}$$

$$y_1' = \frac{1}{6h}(-2y_0 - 3y_1 + 6y_2 - y_3) + \frac{h^3}{12}y^{(4)}$$

$$y_2' = \frac{1}{6h}(y_0 - 6y_1 + 3y_2 + 2y_3) - \frac{h^3}{12}y^{(4)}$$

$$y_3' = \frac{1}{6h}(-2y_0 + 9y_1 - 18y_2 + 11y_3) + \frac{h^3}{4}y^{(4)}$$

n = 4. Five points

$$y_0' = \frac{1}{12h}(-25y_0 + 48y_1 - 36y_2 + 16y_3 - 3y_4) + \frac{h^4}{5}y^{(5)}$$

$$y_1' = \frac{1}{12h}(-3y_0 - 10y_1 + 18y_2 - 6y_3 + y_4) - \frac{h^4}{20}y^{(5)}$$

$$y_2' = \frac{1}{12h}(y_0 - 8y_1 + 8y_3 - y_4) + \frac{h^4}{30}y^{(5)}$$

$$y_3' = \frac{1}{12h}(-y_0 + 6y_1 - 18y_2 + 10y_3 + 3y_4) - \frac{h^4}{20}y^{(5)}$$

$$y_4' = \frac{1}{12h}(3y_0 - 16y_1 + 36y_2 - 48y_3 + 25y_4) + \frac{h^4}{5}y^{(5)}$$

n = 5. Six points

$$y_0' = \frac{1}{60h}(-137y_0 + 300y_1 - 300y_2 + 200y_3 - 75y_4 + 12y_5) - \frac{h^5}{6}y^{(6)}$$

$$y_1' = \frac{1}{60h}(-12y_0 - 65y_1 + 120y_2 - 60y_3 + 20y_4 - 3y_5) + \frac{h^5}{30}y^{(6)}$$

$$y_2' = \frac{1}{60h}(3y_0 - 30y_1 - 20y_2 + 60y_3 - 15y_4 + 2y_5) - \frac{h^5}{60}y^{(6)}$$

$$y_3' = \frac{1}{60h}(-2y_0 + 15y_1 - 60y_2 + 20y_3 + 30y_4 - 3y_5) + \frac{h^5}{60}y^{(6)}$$

(n = 5. Six points)

$$y_4' = \frac{1}{60h}(\ 3y_0 - 20y_1 + 60y_2 - 120y_3.$$
$$+ 65y_4 + 12y_5) - \frac{h^5}{30}y^{(6)}$$

$$y_5' = \frac{1}{60h}(-12y_0 + 75y_1 - 200y_2 + 300y_3$$
$$- 300y_4 + 137y_5) + \frac{h^5}{6}y^{(6)}$$

n = 6. Seven points

$$y_0' = \frac{1}{60h}(-147y_0 + 360y_1 - 450y_2 + 400y_3$$
$$- 225y_4 + 72y_5 - 10y_6) + \frac{h^6}{7}y^{(7)}$$

$$y_1' = \frac{1}{60h}(-10y_0 - 77y_1 + 150y_2 - 100y_3$$
$$+ 50y_4 - 15y_5 + 2y_6) - \frac{h^6}{42}y^{(7)}$$

$$y_2' = \frac{1}{60h}(\ 2y_0 - 24y_1 - 35y_2 + 80y_3$$
$$- 30y_4 + 8y_5 - y_6) + \frac{h^6}{105}y^{(7)}$$

$$y_3' = \frac{1}{60h}(-y_0 + 9y_1 - 45y_2 + 0$$
$$+ 45y_4 - 9y_5 + y_6) - \frac{h^6}{140}y^{(7)}$$

$$y_4' = \frac{1}{60h}(\ y_0 - 8y_1 + 30y_2 - 80y_3$$
$$+ 35y_4 + 24y_5 - 2y_6) + \frac{h^6}{105}y^{(7)}$$

$$y_5' = \frac{1}{60h}(-2y_0 + 15y_1 - 50y_2 + 100y_3$$
$$- 150y_4 + 77y_5 + 10y_6) - \frac{h^6}{42}y^{(7)}$$

$$y_6' = \frac{1}{60h}(\ 10y_0 - 72y_1 + 225y_2 - 400y_3$$
$$+ 450y_4 - 360y_5 + 147y_6) + \frac{h^6}{7}y^{(7)}$$

An examination of these formulas reveals that when the number of points is odd and the derivative is to be found at the mid-point, not only is the formula simpler than others of the same degree but also the error term is smaller. Evidently these formulas are to be chosen whenever a choice is possible. A partial list of these symmetrical formulas is supplied, the subscript notation being modified to exhibit the symmetry.

Formulas for Central Derivatives

$n = 2$

$$y_0' = \frac{1}{2h}(y_1 - y_{-1}) - \frac{h^2}{6}y^{(3)}$$

$n = 4$

$$y_0' = \frac{2}{3h}(y_1 - y_{-1}) - \frac{1}{12h}(y_2 - y_{-2}) + \frac{h^4}{30}y^{(5)}$$

$n = 6$

$$y_0' = \frac{3}{4h}(y_1 - y_{-1}) - \frac{3}{20h}(y_2 - y_{-2}) + \frac{1}{60h}(y_3 - y_{-3}) - \frac{h^6}{140}y^{(7)}$$

$n = 8$

$$y_0' = \frac{4}{5h}(y_1 - y_{-1}) - \frac{1}{5h}(y_2 - y_{-2}) + \frac{4}{105h}(y_3 - y_{-3}) - \frac{1}{280h}(y_4 - y_{-4}) + \frac{h^8}{630}y^{(9)}$$

The formulas for central derivatives are convenient for calculating the derivative from a table of values of the function.

Example 1. Find the derivative of Struve's function $S_0(x)$ at $x = 7.5$.

From a table* of Struve's function we obtain the values

*Jahnke und Emde, p. 220.

x	$S_o(x)$	
7.47	0.1933	
7.48	0.1959	
7.49	0.1984	$y_1 - y_{-1} = 0.0049$
7.50	0.2009	$y_2 - y_{-2} = 0.0099$
7.51	0.2033	$y_3 - y_{-3} = 0.0149$
7.52	0.2058	
7.53	0.2082	

Using these values we have

1) From the 2nd-degree formula $S_o'(7.5) = 0.2450$
2) From the 4th-degree formula $S_o'(7.5) = 0.2442$
3) From the 6th-degree formula $S_o'(7.5) = 0.2438$

Since the differences $y_1 - y_{-1}$, etc., are accurate to two figures only, we cannot hope to obtain the derivative with certainty to more than two places. From the results above we may accept 0.244 as our value of $S_o'(7.5)$ with some uncertainty regarding the last digit.

Exercises

1. From Table VIII at the end of the text calculate $\Gamma'(1)$ using the appropriate four point formula.

2. Find $\Gamma'(1.2)$ using the fourth degree central formula.

3. Calculate $\Psi(.1)$, $\Psi(.2)$, $\Psi(.3)$, etc. by the relation

$$\Psi(x) = \frac{\Gamma'(x + 1)}{\Gamma(x + 1)}$$

and compare with the values tabulated in Table VIII.

4. Find $\Psi'(0.4)$; $\Psi'(0.6)$. Discuss the attainable accuracy for the derivative when the entries for the function are accurate to five places only and h = 0.02.

28. NUMERICAL INTEGRATION

The approximate value of the integral of a function $y = f(x)$, (for which $n + 1$ pairs of values (x_o, y_o), (x_1, y_1), . . ., (x_n, y_n) are known) is found by integrating the interpolating polynomial determined by the given points. Problems of this type arise not only in the case of experimentally determined functions but also in cases where the integrand is explicitly expressed, but the indefinite integral is unknown.

Let $I_k^{(n)}(x)$ denote the integral of Lagrange's coefficient $L_k^{(n)}(x)$ taken from x_0 as a lower limit, so that

$$I_k^{(n)}(x) = \int_{x_0}^{x} L_k^{(n)}(t)\, dt.$$

Then the integral of the given function $y = f(x)$ between the same limits may be expressed as

$$(1) \qquad \int_{x_0}^{x} f(x)\, dx = \sum_{i=0}^{n} I_i^{(n)}(x)\, y_i + E(x)$$

in which E denotes the error committed by using the integral of the interpolating polynomial in place of the true integral. The definite integral between the limits a and b is

$$(2) \qquad \int_a^b f(x)\, dx = \sum_{i=0}^{n} \left[I_i^{(n)}(b) - I_i^{(n)}(a) \right] y_i + E.$$

We notice that the summation on the right is simply a linear combination of the ordinates y_i with coefficients $A_i = \left[I_i^{(n)}(b) - I_i^{(n)}(a) \right]$ which are constants entirely independent of the integrand $f(x)$. Hence, another way of writing equation (2) is

$$(3) \qquad \int_a^b f(x)\, dx = A_0 y_0 + A_1 y_1 + \cdot \cdot \cdot + A_n y_n + E.$$

Now these coefficients A_i can obviously be calculated in any particular case by determining the Lagrangian coefficients for the given set of abscissas $x_0, x_1, \ldots, x_n$ and then integrating between the limits a and b. It is therefore evident that the A_i depend only on the $n + 3$ values of $x, x_0, x_1, \ldots, x_n, a, b$, and are completely determined when these are given. For small values of n we may without difficulty obtain the explicit expressions for the A_i in terms of the given $n + 3$ quantities, but it will be apparent that a <u>general</u> explicit expression, valid for <u>any</u> n, is not easily found.

<u>Example 1</u>. Find the values of the A_i for $n = 2$.

Here we have:

$$L_0^{(2)}(x) = \frac{(x - x_1)(x - x_2)}{(x_0 - x_1)(x_0 - x_2)} = \frac{x^2 - (x_1 + x_2)x + x_1x_2}{(x_0 - x_1)(x_0 - x_2)},$$

$$\int L_0^{(2)}(x)\,dx = \frac{\frac{x^3}{3} - (x_1 + x_2)\frac{x^2}{2} + x_1x_2x}{(x_0 - x_1)(x_0 - x_2)},$$

$$\int_a^b L_0^{(2)}(x)\,dx$$

$$= \frac{2(b^3 - a^3) - 3(x_1 + x_2)(b^2 - a^2) + 6x_1x_2(b - a)}{6(x_0 - x_1)(x_0 - x_2)}$$

whence

$$A_0 = (b - a)\,\frac{2(b^2 + ab + a^2) - 3(x_1 + x_2)(b + a) + 6x_1x_2}{6(x_0 - x_1)(x_0 - x_2)}$$

The values of A_1 and of A_2 can now be found by permutation of the subscripts in the expression for A_0.

$$A_1 = (b - a)\,\frac{2(b^2 + ab + a^2) - 3(x_0 + x_2)(b + a) + 6x_0x_2}{6(x_1 - x_0)(x_1 - x_2)},$$

$$A_2 = (b - a)\frac{2(b^2 + ab + a^2) - 3(x_0 + x_1)(b + a) + 6x_0x_1}{6(x_2 - x_0)(x_2 - x_1)}.$$

Example 2. Estimate $\int_0^{\frac{\pi}{2}} \sin x \, dx$ using the points $x_0 = \frac{\pi}{6}$, $x_1 = \frac{\pi}{4}$, $x_2 = \frac{\pi}{3}$.

Here we have $b = \frac{\pi}{2}$, $a = 0$, and substituting the given values for a, b, x_0, x_1, x_2 in the expressions obtained for A_0, A_1, A_2, in Example 1 we find that

$$A_0 = A_2 = \frac{3\pi}{4}, \quad A_1 = -\frac{4\pi}{4}.$$

Hence by (3)

$$\int_0^{\frac{\pi}{2}} \sin x \, dx = \frac{\pi}{4}\left[3 \sin \frac{\pi}{6} - 4 \sin \frac{\pi}{4} + 3 \sin \frac{\pi}{3}\right] + E$$

This expression, when evaluated, gives:

$$\int_0^{\frac{\pi}{2}} \sin x \, dx = 0.997 + E$$

In this instance the correct value of the integral is known to be 1.000.

Exercises

1. Determine the coefficients A_0, A_1, A_2, A_3 for the case where $n = 3$.
2. Letting $x_0 = x_0$, $x_1 = x_0 + h$, $x_2 = x_0 + 2h$, $x_3 = x_0 + 3h$, $a = x_0$, $b = x_3$ in Exercise 1, obtain the formula:

$$\int_{x_0}^{x_0+3h} f(x) \, dx = \frac{3h}{8}\left[y_0 + 3y_1 + 3y_2 + y_3\right] + E.$$

3. Letting $x_0 = x_0$, $x_1 = x_0 + h$, $x_2 = x_0 + 2h$, $x_3 = x_0 + 3h$, $a = x_0$, $b = x_0 + 2h$ in Exercise 1, show that

$$\int_{x_0}^{x_0+2h} f(x)\,dx = \frac{h}{3}\Big[y_0 + 4y_1 + y_2\Big] + E.$$

Note that the coefficient of y_3 vanishes.

29. UNDETERMINED COEFFICIENTS

We have seen how formulas for numerical differentiation and integration are obtainable by differentiating or integrating the interpolating polynomial. The same results can be secured by the use of undetermined coefficients, a method which not only is frequently simpler in application but which also can be used for the derivation of more general classes of formulas than those resulting from the differentiation or integration of interpolating polynomials.

Let us first illustrate the method of undetermined coefficients by some examples.

Example 1. Find a formula for dy/dx at $x = x_1$ in terms of ordinates y_0, y_1, y_2, y_3, at the four equally-spaced values $x_0, x_0 + h, x_0 + 2h, x_0 + 3h$. Using undetermined coefficients the desired formula may be expressed as

$$y_1' = C_0y_0 + C_1y_1 + C_2y_2 + C_3y_3. \tag{1}$$

To determine the coefficients we impose the condition that the formula is to be exact whenever y is any polynomial of degree not more than 3. In particular then the formula is to be exact if $y = 1$, if $y = (x - x_0)$, if $y = (x - x_0)^2$, or if $y = (x - x_0)^3$. When we substitute in turn these four values of y into equation (1) there results four linear equations which the coefficients C_i must satisfy:

$$\begin{aligned}
0 &= C_0 + C_1 + C_2 + C_3, \\
1 &= hC_1 + 2hC_2 + 3hC_3, \\
2h &= h^2C_1 + 4h^2C_2 + 9h^2C_3, \\
3h^2 &= h^3C_1 + 8h^3C_2 + 27h^3C_3.
\end{aligned}$$

Upon solving for the C's and putting the resulting values back in (1) we get

$$y_1' = \frac{1}{6h}(-2y_0 - 3y_1 + 6y_2 - y_3).$$

This is seen to agree with the second formula for the case of 4 points on p. 97.

Example 2. Express the integral from x_1 to x_2 in terms of the four ordinates at the four equally-spaced points x_0, x_1, x_2, x_3.

The desired formula is

$$\int_{x_1}^{x_2} y\,dx = A_0y_0 + A_1y_1 + A_2y_2 + A_3y_3.$$

In order to avail ourselves of the symmetry present in this example, we set $x_0 = a - \frac{3h}{2}$, $x_1 = a - \frac{h}{2}$, $x_2 = a + \frac{h}{2}$, $x_3 = a + \frac{3h}{2}$, and letting $y = 1$, $y = (x - a)$, $y = (x - a)^2$, $y = (x - a)^3$ in turn we get the four equations (after removal of the factors $\frac{h}{2}$, $\frac{h^2}{4}$, $\frac{h^3}{8}$)

$$\begin{aligned}
h &= A_0 + A_1 + A_2 + A_3, \\
0 &= -3A_0 - A_1 + A_2 + 3A_3, \\
\frac{h}{3} &= 9A_0 + A_1 + A_2 + 9A_3, \\
0 &= -27A_0 - A_1 + A_2 + 27A_3.
\end{aligned}$$

We find that

$$A_0 = A_3 = -\frac{h}{24}$$

$$A_1 = A_2 = \frac{13h}{24}$$

and the final formula is therefore

$$\int_{x_1}^{x_2} y\,dx = \frac{h}{24}\left[-y_0 + 13y_1 + 13y_2 - y_3\right].$$

As an instance of a formula of more general type, we consider the following.

Example 3. Express the integral from x_0 to x_2 in terms of the three ordinates and three derivatives at the three equally-spaced points x_0, x_1, x_2.

The required formula is of the type

$$\int_{x_0}^{x_2} y\,dx = A_0y_0 + A_1y_1 + A_2y_2 + B_0y_0' + B_1y_1' + B_2y_2'.$$

Here we set $x_0 = x_1 - h$, $x_2 = x_1 + h$, and let $y = (x - x_1)^k$, $k = 0, 1, \ldots, 5$. The six equations that result are, after obvious factors have been removed:

$$\begin{aligned}
2h &= A_0 + A_1 + A_2,\\
0 &= -A_0h + A_2h + B_0 + B_1 + B_2,\\
\frac{2h^2}{3} &= A_0h + A_2h - 2B_0 + 2B_2,\\
0 &= -A_0h + A_2h + 3B_0 + 3B_2,\\
\frac{2h^2}{5} &= A_0h + A_2h - 4B_0 + 4B_2,\\
0 &= -A_0h + A_2h + 5B_0 + 5B_2,
\end{aligned}$$

From the second, fourth, and sixth equations it follows

that

$$A_0 = A_2, \qquad B_0 = -B_2, \qquad B_1 = 0.$$

The remaining three equations are thereby reduced to

$$\begin{aligned} 2h &= 2A_0 + A_1 \\ \frac{2h}{3} &= 2A_0 \qquad - 4B_0 \\ \frac{2h}{5} &= 2A_0 \qquad - 8B_0 \end{aligned}$$

whence $B_0 = \frac{h^2}{15}$, $A_0 = \frac{7h}{15}$, $A_1 = \frac{16h}{15}$, and the final formula is

$$\int_{x_0}^{x_2} y\,dx = \frac{h}{15}(7y_0 + 16y_1 + 7y_2) + \frac{h^2}{15}(y_0' - y_2').$$

Exercises

Using undetermined coefficients establish the following formulas and show that each is exact if y is a polynomial of degree indicated.

1. $y_2 = y_{-2} + \frac{4h}{3}(2y_1' - y_0' + 2y_{-1}')$ (4th degree)

2. $y_1 = y_0 + \frac{h}{24}(7y_1' + 16y_0' + y_{-1}') + \frac{h^2y_0''}{4}$ (4th degree)

3. $y_2 = y_0 + \frac{2h}{3}(5y_1' - y_0' - y_{-1}') - 2h^2y_0''$ (4th degree)

4. $y_2 - 2y_1 + y_0 = \frac{h^2}{12}(y_2'' + 10y_1'' + y_0'')$ (5th degree)

5. $y_4 - y_3 - y_1 + y_0 = \frac{h^2}{4}(5y_3'' + 2y_2'' + 5y_1'')$ (5th degree)

6. $y_1 = y_0 + \frac{h}{2}(y_1' + y_0') - \frac{h^2}{12}(y_1'' - y_0'')$ (4th degree)

7. $y_1 - 2y_0 + y_{-1} = \frac{3h}{8}(y_1' - y_{-1}')$
$\qquad - \frac{h^2}{24}(y_1'' - 8y_0'' + y_{-1}'')$ (7th degree)

8. $y_5 - y_0 = \frac{5h}{24}(11y_4' + y_3' + y_2' + 11y_1')$ (4th degree)

30. INVESTIGATION OF THE ERROR

The formulas derived in the exercises and examples of the preceding articles may be looked upon as the result of applying an operator to the given function. For instance in the third example of Art. 29 if we replace y by the function f(x) and add a term R(f) to designate the error of the formula, we see that

$$\int_{x_0}^{x_2} f(x)\,dx = \frac{h}{15}\left[7f(x_2) + 16f(x_1) + 7f(x_0)\right] + \frac{h^2}{15}\left[f'(x_0) - f'(x_2)\right] + R(f).$$

From this follows

$$R(f) = \int_{x_0}^{x_2} f(x)\,dx - \frac{h}{15}\left[7f(x_2) + 16f(x_1) + 7f(x_0)\right] + \frac{h^2}{15}\left[f'(x_2) - f'(x_0)\right].$$

The right-hand member of this equation may be considered as the result of performing on the function f(x) an operation denoted by the symbol R(f).

It is clear that this operator is a linear operator, for if $f(x) = a\,u(x) + bv(x)$, where a and b are any constants, then

$$R(f) = aR(u) + bR(v).$$

The operation R(f) also has the property of reducing identically to zero if f(x) is replaced by any polynomial of degree not exceeding a definite value n depending on R. In the example cited above the formula is exact for all polynomials of degree 5 or less but is not exact for $f(x) = x^6$, so that $R(f) = 0$ if $f = 1, x, \ldots, x^5$, but $R(f) \neq 0$ if $f = x^6$. It is convenient to say that an operator R is of <u>degree n</u> when $R(x^m) = 0$ for $m \leqq n$ but $R(x^{n+1}) \neq 0$. This concept of an operator is quite general and applies not only to all the formulas of this chapter but to the interpolation formulas of Chapter III as well. The idea may also be extended to the case where the approximating functions are other than polynomials but that possibility will not be stressed here.

Our immediate purpose in introducing the operator R is to derive a general form for the error involved in any formula of the type $R(f) = 0$. When $R(f)$ is not actually zero and we use the formula as though $R(f)$ were zero the error committed is evidently the value of $R(f)$. We now proceed to obtain an expression from which the magnitude of $R(f)$ can be estimated.

Let $f(s)$ be a function with a continuous derivative of order $(n+1)$ in the interval $a \leqq s \leqq x$. Then repeated integrations by parts give

$$\frac{1}{n!}\int_a^x f^{(n+1)}(s)(x - s)^n ds = -\frac{f^{(n)}(a)(x - a)^n}{n!} - \cdots - \frac{f''(a)(x - a)^2}{2!} - f'(a)(x - a) - f(a) + f(x).$$

Solving this equation for $f(x)$ we get

$$(1)\quad f(x) = f(a) + f'(a)(x - a) + \frac{f''(a)}{2!}(x - a)^2 + \cdots + \frac{f^{(n)}(a)}{n!}(x - a)^n + \frac{1}{n!}\int_a^x f^{(n+1)}(s)(x - s)^n ds.$$

The right-hand member of (1) is seen to be the first $n + 1$ terms of Taylor's series with a remainder expressed as an integral. We note that the first $n + 1$ terms on the right form a polynomial in x of degree n which we may call $Q_n(x)$ so that (1) can be written

$$(2) \qquad f(x) = Q_n(x) + \frac{1}{n!} \int_a^x f^{(n+1)}(s)(x - s)^n ds.$$

It will be convenient to introduce the notation $\overline{(x - s)}^n$ by the definition

$$\overline{(x - s)}^n = (x - s)^n \quad \text{if } x > s,$$

$$\overline{(x - s)}^n = 0 \quad \text{if } x < s.$$

In particular

$$\overline{(x - s)}^0 = 1 \quad \text{if } x > s,$$

$$\overline{(x - s)}^0 = 0 \quad \text{if } x < s.$$

Also

$$\frac{d}{dx}\overline{(x - s)}^n = n\overline{(x - s)}^{n-1},$$

and

$$\int_a^b \overline{(x - s)}^n dx = \frac{\overline{(b - s)}^{n+1} - \overline{(a - s)}^{n+1}}{n + 1}$$

Using this notation we may write (2) in the form

$$(3) \qquad f(x) = Q_n(x) + \frac{1}{n!} \int_a^{\infty} f^{(n+1)}(s)\overline{(x - s)}^n ds$$

because the integrand is zero for all values of s greater than x.

Now let R be any linear operator of degree n, which does not involve derivatives of f(x) of order exceeding n - 1. Also let the constant <u>a</u> in (3) be chosen so as to be less than the least value of x used in performing the operation R(f). If we apply the operator R to both members of equation (3) we shall have

$$(4) \qquad R(f) = R(Q_n) + \frac{1}{n!} \int_a^\infty f^{(n+1)}(s)\, R_x\, (\overline{x - s})^n\, ds$$

in which $R_x\left[(\overline{x - s})^n\right]$ means the operator R applied to $(\overline{x - s})^n$ regarded as a function of the variable x. Since $Q_n(x)$ is a polynomial of degree n it follows from our hypotheses that $R(Q_n) = 0$, and equation (4) becomes

$$R(f) = \int_a^\infty f^{(n+1)}(s)\, G(s)\, ds,$$

where

$$G(s) = \frac{1}{n!} R_x \left[(\overline{x - s})^n\right].$$

Whenever s is less than the least x occurring in R the function G(s) vanishes identically, for then the terms x - s will all be positive so that $(\overline{x - s})^n = (x - s)^n$, and

$$G(s) = \frac{1}{n!} R_x \left[(x - s)^n\right] = 0$$

since $(x - s)^n$ is a polynomial of degree n. For this reason the lower limit <u>a</u> in the integral may be replaced by $-\infty$, and we have finally

$$(5) \qquad R(f) = \int_{-\infty}^{\infty} f^{(n+1)}(s)\, G(s)\, ds.$$

Whenever s is greater than the greatest x in R the function G again vanishes identically, for in this case all the terms in R $\left[\overline{(x - s)}^n\right]$ are identically zero by the definition of $\overline{(x - s)}^n$. It should be noted here that we have excluded the possibility of a term $\overline{(x - s)}^0$ occurring, since we have assumed that the operator R involves no derivative of order exceeding $n - 1$.

In the interval between the least and greatest x involved in R, the function G(s) consists of a series of polynomial arcs joined together continuously. This may be illustrated by forming G(s) for the operator

$$R(f) = \int_{x_0}^{x_2} f(x)\, dx - \frac{h}{15}\left[7f(x_2) + 16f(x_1) + 7f(x_0)\right] + \frac{h^2}{15}\left[f'(x_2) - f'(x_0)\right]$$

If we let $x_0 = -h$, $x_1 = 0$, $x_2 = h$ and note that the degree of this operator is 5, we see that

$$G(s) = \frac{1}{5!}\left[\frac{\overline{(h - s)}^6 - \overline{(-h - s)}^6}{6} - \frac{h}{15}\left[7\overline{(h - s)}^5 + 16\overline{(-s)}^5 + 7\overline{(-h - s)}^5\right] + \frac{h^2}{3}\left[\overline{(h - s)}^4 - \overline{(-h - s)}^4\right]\right]$$

If $s > 0$, this reduces to

$$G(s) = \frac{(h - s)^4}{3600}\,(5s^2 + 4hs + h^2)$$

On the other hand, if $s < 0$

$$G(s) = \frac{1}{3600}(h + s)^4(5s^2 - 4hs + h^2)$$

The graph of $G(s)$ has the form shown below:

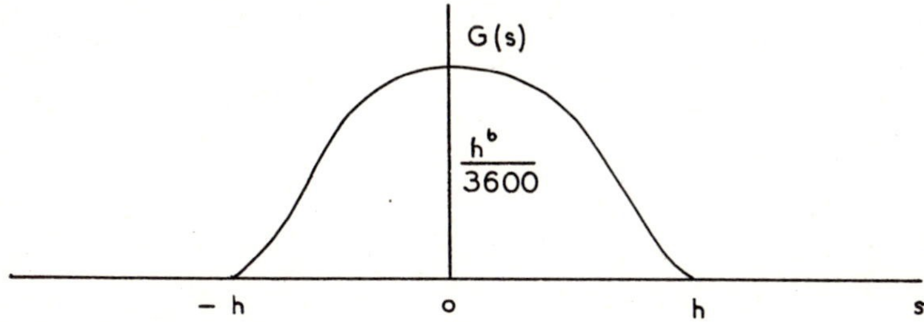

The error of the given formula is by (5)

$$R(s) = \int_{-\infty}^{\infty} f^{(6)}(s)\, G(s)\, ds.$$

Since the function $G(s)$ does not change sign the law of the mean gives

$$R(f) = f^{(6)}(z) \int_{-h}^{h} G(s)\, ds$$

where z is some value of s between $-h$ and h. Since $G(s)$ is explicitly known in each subinterval, the integral can be evaluated and the final form of the error term is

$$R(f) = \frac{f^{(6)}(z)h^7}{4725}$$

Exercises

Obtain the function $G(s)$ for each of the formulas in the exercises of Art. 29.

31. CALCULATION OF THE ERROR

In formula (5) of Art. 30, the error was expressed in the form

$$R(f) = \int_{-\infty}^{\infty} f^{(n+1)}(s)\, G(s)\, ds.$$

From this relation, we may conclude that

$$|R(f)| \leqq \max |f^{(n+1)}(s)| \int_{-\infty}^{\infty} |G(s)|\, ds$$

where $\max |f^{(n+1)}(s)|$ means the maximum of $f^{(n+1)}(x)$ in the range of values of x involved in the operator R. Thus as a general result we have:

Theorem 1. _If_ R _is any linear operator of degree_ n, _not involving differentiations of order greater than_ n - 1, _and if_ f(x) _is any function having a derivative of order_ n + 1 _in the range covered by_ R, _the error_ R(f) _is bounded by the inequality_

$$|R(f)| < \max |f^{(n+1)}(s)|\, K$$

where $K = \int_{-\infty}^{\infty} |G(s)|\, ds$ _and is entirely independent of the function_ f(x).

It turns out that for many of the most important formulas used in numerical interpolation, differentiation, integration, and for the solution of differential equations the function G(s) associated with the formula does not change sign in the interval $-\infty$ to ∞. This was seen to be the case in the example treated in Art. 30. Since the function $G(s)$ depends on the operator only and not on the function operated upon, it is appropriate to introduce the term <u>definite operator</u> for an operator whose $G(s)$ does not change sign, and the term <u>indefinite operator</u> for the case where $G(s)$ changes sign. The analogy with definite and indefinite quadratic forms suggests the use of these terms.

Theorem 2. *If the operator R is definite then*

$$R(f) = f^{(n+1)}(z)\, R\left(\frac{x^{n+1}}{(n+1)!}\right),$$

where z lies between the greatest and the least values of x involved in R.

Since by hypothesis $G(s)$ does not change sign the law of the mean gives

$$R(f) = \int_{-\infty}^{\infty} f^{(n+1)}(s)\, G(s)\, ds = f^{(n+1)}(z) \int_{-\infty}^{\infty} G(s)\, ds.$$

In particular if $f(x) = \frac{x^{n+1}}{(n+1)!}$, and consequently $f^{(n+1)}(z) = 1$, the above equation becomes

$$R\left(\frac{x^{n+1}}{(n+1)!}\right) = \int_{-\infty}^{\infty} G(s)\, ds,$$

so that

$$R(f) = f^{(n+1)}(z)\, R\left[\frac{x^{n+1}}{(n+1)!}\right]$$

The practical value of this result lies in the fact that it is much easier to substitute $\frac{x^{n+1}}{(n+1)!}$ into the given formula than it is to compute the polynomial segments of which $G(s)$ is composed and then perform the integrations necessary to obtain $\int_{-\infty}^{\infty} G(s)\, ds$. There remains of course the difficulty that we may not be able to assure ourselves that R is definite without first calculating G(s). Even so it is usually easier to substitute $x^{n+1}/(n + 1)!$ than to evaluate $\int_{-\infty}^{\infty} G(s)\, ds$.

In the example of Art. 30

$$R(f) = \int_{-h}^{h} f(x)\, dx - \frac{h}{15}\Big[7f(h) + 16f(0) + 7f(-h)\Big] + \frac{h^2}{15}\Big[f'(h) - f'(-h)\Big].$$

Setting $f(x) = x^6/6!$ we have

$$R(x^6/6!) = \frac{1}{6!}\left[\frac{2h^7}{7} - \frac{2\cdot 7h^7}{15} + \frac{2\cdot 6h^7}{15}\right] = \frac{h^7}{4725}$$

and the error term is

$$R(f) = \frac{f^{(6)}(z)h^7}{4725},$$

as had already been shown by the other method.

Exercises

Determine the error terms in the exercises of Art. 29.

32. THE TRAPEZOIDAL RULE

If $y_0, y_1, y_2, \ldots,$ are the values of $y = f(x)$ at equally-spaced points $x_0, x_1, x_2, \ldots,$ with interval h the method of undetermined coefficients gives the formula

$$(1) \qquad \int_{x_{n-1}}^{x_n} y\, dx = \frac{h}{2}(y_{n-1} + y_n) - \frac{y''h^3}{12},$$

where the error is determined as in Art. 31. Setting $n = 1, 2, \ldots, m$ in succession and adding we have

$$(2) \qquad \int_{x_0}^{x_m} y\,dx$$

$$= h(\tfrac{1}{2}y_0 + y_1 + y_2 + \cdots + y_{m-1} + \tfrac{1}{2}y_m) + E.$$

The error E may be shown to be

$$E = -\frac{my''h^3}{12}$$

in which y" denotes the second derivative of y taken at some point between x_0 and x_m. Formula (2) is the well-known trapezoidal rule for the evaluation of an integral. In many respects it is the simplest and most adaptable formula for numerical integration, its principal defect being that its accuracy is not of a high order.

Example 1. Find the integral of cos x from $x = 0$ to $x = 0.8$ using $h = 0.1$.

The required values of cos x are tabulated below:

x	cos x
0	1.00000000
0.1	0.99500417
0.2	0.98006658
0.3	0.95533649
0.4	0.92106099
0.5	0.87758256
0.6	0.82533562
0.7	0.76484219
0.8	0.69670671

The value of the integral, as given by (1), is easily obtained and proves to be 0.716758196. If, however, we calculate E, using the fact that $|y''| \leq 1$ for the interval in question, we find that

$$E < \frac{8(0.1)^3}{12} = 0.00067,$$

which casts doubt on the last six figures of our result. In actual fact, the error is about 0.00059789.

In spite of the size of the error, the trapezoidal rule is very useful provided its limitations are observed.

By the addition of a relatively simple correction term the accuracy of the trapezoidal rule can be materially improved. For this purpose, we use the formula

$$(3) \qquad \int_{x_{n-1}}^{x_n} y\,dx$$

$$= \frac{h}{24}\left[-y_{n-2} + 13y_{n-1} + 13y_n - y_{n+1}\right] + \frac{11y^{(4)}h^5}{720}$$

which is found in the usual manner by undetermined coefficients. If we let $n = 1, 2, \ldots, m$ and add, we derive the equation

$$(4) \qquad \int_{x_0}^{x_m} y\,dx = h\left[\tfrac{1}{2}y_0 + y_1 + y_2 + \cdots + y_{m-1} + \tfrac{1}{2}y_m\right]$$

$$+ \frac{h}{24}\left[-y_{-1} + y_1 + y_{m-1} - y_{m+1}\right] + \frac{11my^{(4)}h^5}{720}$$

The second term on the right in this equation is the correction term. It depends, rather surprisingly, on four ordinates only, regardless of m. It requires the knowledge of two ordinates, y_{-1} and y_{m+1}, outside the range of integration, which is sometimes a handicap, but this is usually more than compensated for by the increase in accuracy.

Example 2. Work Example 1 by means of formula (4). Here we need the values

x	cos x
-0.1	0.99500417
0.1	0.99500417
- - -	- - -
0.7	0.76484219
0.9	0.62160997

From these the correction term is found to be 0.000596801, which, combined with the result found in Example 1, gives the answer 0.717354997. The maximum value of the error term is 0.00000123, and the actual error is 0.00000109. Since the error of the uncorrected trapezoidal rule was 0.00059789, the addition of the correction term has secured a marked increase in accuracy.

In a similar manner by starting from the formula

$$\int_{x_{n-1}}^{x_n} y\,dx = \frac{h}{1440}\left[11y_{n-3} - 93y_{n-2} + 802y_{n-1} + 802y_n - 93y_{n+1} + 11y_{n+2}\right]$$

we may obtain the trapezoidal rule with two correction terms

$$(5) \qquad \int_{x_0}^{x_m} y\,dx = h\left[\tfrac{1}{2}y_0 + y_1 + y_2 + \cdot\cdot\cdot + y_{m-1} + \tfrac{1}{2}y_m\right] + \frac{h}{24}\left[-y_{-1} + y_1 + y_{m-1} - y_{m+1}\right] + \frac{11h}{1440}\left[(y_{-2} - 2y_{-1} + 2y_1 - y_2) + (-y_{m-2} + 2y_{m-1} - 2y_{m+1} + y_{m+2})\right] - \frac{191my^{(6)}h^7}{60480}.$$

As a matter of fact, any number of formulas of this type can be obtained each with one more correction term than the preceding. All are special cases of Gregory's formula, which itself is a modification of the Euler-Maclaurin summation formula. (See Chapter VI.)

Example 3. Work Example 1 by means of formula (5).

The second correction term is found to be 0.000001093, and this, when added to the result of Example 2, gives 0.717356090. This answer is in error by one unit in the 9th decimal place.

In comparison with most of the formulas for numerical integration which will be taken up later those given here enjoy one marked advantage. They can be applied to any number of intervals without restriction. Many of the later formulas apply only if the number of intervals is a multiple of 2, or a multiple of 3, or a multiple of 4, etc.

Exercises

1. From the integral $\ln x = \int_1^x \frac{dt}{t}$, calculate ln x for x = 1.1, 1.2, 1.3, 1.4, 1.5 using h = 0.1. Compare the actual error with the theoretical error. Use formula (2) of this Article.

2. Work Exercise 1 by means of formula (3).

3. Using formula (2) calculate π from the integral

$$\pi = 4 \int_0^1 \frac{dx}{1 + x^2}.$$

Choose h so as to be assured of accuracy to four decimal places.

33. SIMPSON'S RULE

If $y_0, y_1, y_2, \ldots,$ are the values of $y = f(x)$ at equally-spaced points $x_0, x_1, x_2, \ldots,$ with interval h the method of undetermined coefficients gives the equation

$$(1) \quad \int_{x_{n-1}}^{x_{n+1}} y\,dx = \frac{h}{3}(y_{n-1} + 4y_n + y_{n+1}) - \frac{y^{(4)}h^5}{90}$$

This formula is noteworthy because its accuracy is actually greater than might be expected of a formula containing only three ordinates. For in general with only three coefficients to be determined, we would expect to get a formula that is exact for all polynomials of degree two but not of degree three. Formula (1), however, is exact if y is any polynomial of degree three.

If in (1) we set $n = 1, 3, 5, 7, \ldots, 2m - 1$, and add the resulting equations, we are led to Simpson's Rule:

$$(2) \quad \int_{x_0}^{x_{2m}} y\,dx = \frac{h}{3}(y_0 + 4y_1 + 2y_2 + 4y_3 + 2y_4 + 4y_5 + \cdots + 4y_{2m-1} + y_{2m}) - \frac{my^{(4)}h^5}{90}$$

In the remainder term $y^{(4)}$ denotes the value of the fourth derivative at some point between x_0 and x_{2m}.

Example. Find the integral of cos x from 0 to 0.8 by Simpson's Rule, using $h = 0.1$.

The values of cos x and the corresponding multipliers in Simpson's Rule are tabulated below:

x	cos x	Multiplier
0	1.00000000	1
0.1	0.99500417	4
0.2	0.98006658	2
0.3	0.95533649	4
0.4	0.92106099	2
0.5	0.87758256	4

0.6	0.82533562	2
0.7	0.76484219	4
0.8	0.69670671	1

The multiplication and summation is performed in a continuous operation on the calculating machine, and the result is then multiplied by $h/3$. In this manner we find the integral to be 0.71735649. The error term when evaluated with $y^{(4)} = 1$ has the value -0.00000045. The actual error is -0.00000040.

Simpson's Rule combines simplicity of form with rather good accuracy, and consequently is much used for numerical integration. It can be applied only to an <u>even</u> number of intervals, which in some situations is a serious inconvenience. For instance, in the example given above, it would be impracticable to obtain the integral from 0 to 0.7 by Simpson's Rule. The formulas of Article 32, however, would apply without restriction.

Exercises

1. Evaluate $\int_0^1 \frac{dx}{1 + x^4}$, $h = 0.1$.

2. Evaluate $\int_0^{1.2} \sqrt{1 + x^3}\, dx$, $h = 0.2$.

3. Evaluate $\int_1^3 \frac{dx}{x}$, $h = 0.5$.

34. NEWTON-COTES QUADRATURE FORMULAS, CLOSED TYPE

Formula (1) of Art. 32 and (1) of Art. 33 are the first two in a sequence of formulas of the type

$$\int_{x_0}^{x_n} y\, dx = A_0 y_0 + A_1 y_1 + \cdot \cdot \cdot + A_2 y_n.$$

Each of these can be obtained by means of undetermined coefficients, and the error is computed in the usual manner. The first eight are listed below:

(1) $$\int_{x_0}^{x_1} y\,dx = \frac{h}{2}(y_0 + y_1) - \frac{y''h^3}{12}.$$

(2) $$\int_{x_0}^{x_2} y\,dx = \frac{h}{3}(y_0 + 4y_1 + y_2) - \frac{y^{(4)}h^5}{90}.$$

(3) $$\int_{x_0}^{x_3} y\,dx = \frac{3h}{8}(y_0 + 3y_1 + 3y_2 + y_3) - \frac{3y^{(4)}h^5}{80}.$$

(4) $$\int_{x_0}^{x_4} y\,dx = \frac{4h}{90}(7y_0 + 32y_1 + 12y_2 + 32y_3 + 7y_4) - \frac{8y^{(6)}h^7}{945}.$$

(5) $$\int_{x_0}^{x_5} y\,dx = \frac{5h}{288}(19y_0 + 75y_1 + 50y_2 + 50y_3 + 75y_4 + 19y_5) - \frac{275y^{(6)}h^7}{12096}$$

(6) $$\int_{x_0}^{x_6} y\,dx = \frac{6h}{840}(41y_0 + 216y_1 + 27y_2 + 272y_3 + 27y_4 + 216y_5 + 41y_6) - \frac{9y^{(8)}h^9}{1400}.$$

$$(7) \quad \int_{x_0}^{x_7} y\,dx = \frac{7h}{17280}(751y_0 + 3577y_1 + 1323y_2 + 2989y_3$$

$$+ 2989y_4 + 1323y_5 + 3577y_6 + 751y_7) - \frac{8183y^{(8)}h^9}{518400}.$$

$$(8) \quad \int_{x_0}^{x_8} y\,dx = \frac{8h}{28350}(989y_0 + 5888y_1 - 928y_2 + 10496y_3$$

$$- 4540y_4 + 10496y_5 - 928y_6 + 5888y_7 + 989y_8) - \frac{2368y^{(10)}h^{1}}{467775}$$

An examination of the error terms reveals that (2), (4), (6), and (8) are exact for all polynomials of degree equal to the number of ordinates used, while (1), (3), (5), and (7) are exact only for polynomials of degree one less than the number of ordinates. Hence, it turns out that (2) and (3) are of the same order of accuracy although (3) has one more ordinate. The same is true of the pair (4) and (5) and the pair (6) and (7), etc. In general, therefore, when we have any choice in the matter, it is better to use an even-numbered formula rather than the corresponding odd-numbered formula, since we get approximately the same accuracy with one less ordinate to compute. This does not mean that odd-numbered formulas should not be used when occasion requires it. For instance, if we need to obtain an integral over eleven intervals with the approximate accuracy of Simpson's Rule, we may use (2) four times and (3) once.

Formula (3) is commonly known as Newton's "Three-eighths rule." When Newton's Three-eights rule is used with the same interval h as Simpson's rule it is clear that Simpson is more accurate, the respective coefficients being $\frac{3}{80}$ and $\frac{1}{90}$. If, however, the two be compared for the same total interval L of integration the h in Simpson's rule is L/2, the h in Newton's is L/3, hence the respective errors are $-y^{(4)}L^5/2880$ and $-y^{(4)}L^5/6480$, so that in this case Newton has the better of it.

While the foregoing formulas increase in accuracy as the number of terms increases the complexity of the coefficients also increases and the adaptability to fit a given number of intervals decreases. Hence, the selection of a suitable formula in a given problem is usually a compromise between accuracy on the one hand and convenience on the other.

Weddle's Rule. Equation (6) may be written the form

$$\int_{x_0}^{x_6} y\,dx = \frac{6h}{840}\Big[(42-1)y_0 + (210+6)y_1 + (42-15)y_2$$

$$+ 252+20)y_3 + (42-15)y_4 + (210+6)y_5 + (42-1)y_6\Big] + E$$

$$= \frac{3h}{10}(y_0 + 5y_1 + y_2 + 6y_3 + y_4 + 5y_5 + y_6)$$

$$-\frac{h}{140}(y_0 - 6y_1 + 15y_2 - 20y_3 + 15y_4 - 6y_5 + y_6) + E.$$

The operator R, where

$$R(y) = y_0 - 6y_1 + 15y_2 - 20y_3 + 15y_4 - 6y_5 + y_6$$

is of degree five and by means of formula (5), Art. 30, we find that

$$R(y) = y^{(6)}h^6.$$

Hence

$$(9)\qquad \int_{x_0}^{x_6} y\,dx = \frac{3h}{10}(y_0 + 5y_1 + y_2 + 6y_3 + y_4 + 5y_5 + y_6)$$

$$-\frac{y^{(6)}h^7}{140} - \frac{9y^{(8)}h^9}{1400}.$$

Formula (9) is known as "Weddle's Rule." While its order of accuracy is less than that of (6) the simplicity of its coefficients makes it a popular formula for numerical integration.

Exercises

1. Calculate $\ln 1.4 = \int_1^{1.4} dx/x$ by (4), $h = 0.1$.

2. Calculate $\ln 1.6$ by Weddle's rule, $h = 0.1$.

3. Calculate $\ln 2$ by (5), $h = 0.2$.

35. NEWTON-COTES QUADRATURE FORMULAS, OPEN TYPE

The set of formulas given in the preceding article are all of the type

$$\int_{x_o}^{x_n} y\, dx = A_o y_o + A_1 y_1 + \cdots + A_n y_n.$$

A set of similar formulas of the type

$$\int_{x_o}^{x_n} y\, dx = A_1 y_1 + A_2 y_2 + \cdots + A_{n-1}\, y_{n-1},$$

in which the ordinates at the ends of the interval of integration do not occur, can be obtained in the same manner. These are called "open" formulas while those of Article 34 are said to be "closed". The principal use of the open formulas is in the numerical solution of differential equations. The first six are listed below:

$$\int_{x_0}^{x_3} y\,dx = \frac{3h}{2}(y_1 + y_2) + \frac{y''h^3}{4}. \tag{1}$$

$$\int_{x_0}^{x_4} y\,dx = \frac{4h}{3}(2y_1 - y_2 + 2y_3) + \frac{28y^{(4)}h^5}{90}. \tag{2}$$

$$\int_{x_0}^{x_5} y\,dx = \frac{5h}{24}(11y_1 + y_2 + y_3 + 11y_4) + \frac{95y^{(4)}h^5}{144}. \tag{3}$$

$$\int_{x_0}^{x_6} y\,dx = \frac{6h}{20}(11y_1 - 14y_2 + 26y_3 - 14y_4 + 11y_5) + \frac{41y^{(6)}h^7}{140}. \tag{4}$$

$$\int_{x_0}^{x_7} y\,dx = \frac{7h}{1440}(611y_1 - 453y_2 + 562y_3 + 562y_4 - 453y_5 + 611y_6) + \frac{5257}{8640}y^{(6)}h^7. \tag{5}$$

$$\int_{x_0}^{x_8} y\,dx = \frac{8h}{945}(460y_1 - 954y_2 + 2196y_3 - 2459y_4 + 2196y_5 - 954y_6 + 460y_7) + \frac{3956}{14175}y^{(8)}h^9. \tag{6}$$

36. SUMMARY

In this chapter we have shown how a wide variety of formulas can be obtained by the method of undetermined coefficients and how an expression for the error is secured. We have also listed a considerable number of the specific formulas most used for numerical differentiation and integration. All these explicit formulas are subject to the rather severe limitation that the points used are equally spaced and that the derivative or integral as the case may be is evaluated at the given points and not for some intermediate value of x. In problems where these conditions are not fulfilled, it is probably best to obtain the interpolating polynomial and then differentiate or integrate as may be required.

One further difficulty should be discussed. In all cases our error is made to depend on $f^{(n+1)}(x)$, but in practical problems it is often difficult or even impossible to calculate $f^{(n+1)}(x)$. Under these conditions what can be done about estimating the error?

If nothing whatever is known about the function $y = f(x)$ except the given pairs of values (x_0, y_0), (x_1, y_1), . . ., then indeed nothing whatever can be known about the error. But if we are justified in assuming that $f^{(n+1)}(x)$, though unknown, does not change rapidly in value in the range of x considered, the following formula is useful

$$(1) \quad h^{(n+1)}f^{(n+1)}(z) = y_{n+1} - \binom{n+1}{1} y_n + \binom{n+1}{2} y_{n-1} - \binom{n+1}{3} y_{n-2} + \cdots$$

where $\binom{n+1}{1}$, $\binom{n+1}{2}$, etc. are the binomial coefficients belonging to the exponent $n+1$. Here the y's are the ordinates at equally-spaced values of x with the interval h, and z is some value between x_0 and x_n. The expression on the right is easily remembered because of its analogy to the binomial expansion. In fact if we expand $(y - 1)^{n+1}$ by the binomial theorem, rewrite the exponents of the y's as subscripts, and replace the final 1 by y_0, we obtain exactly the right-hand member of (1). Now of course the z here will not be the z occurring in the expression for the error, but under our assumption that $f^{(n+1)}(z)$ does not change rapidly we may use the value thus obtained as an approximation to that needed for the error.

Example. Estimate the value of $f^{(4)}(z)$ from the data of Example 1, p. 117.

Here we have as multipliers the binomial coefficients 1, -4, 6, -4, 1, and applying these multipliers to the first five values we get 0.00009783. Since $h = 0.1$, this gives $f^{(4)}(z) = 0.9783$. Actually, of course, $f^{(4)}(x) = \cos x$, and its value in the range from $x = 0$ to $x = 0.4$ varies from 1.0000 to 0.9211.

Additional formulas for numerical integration will be found in Chapters VI and IX.

Exercises

1. Calculate $\int_0^{12} x^7 dx$ by (4), Article 34, using $h = 1$.

   Also by (6), Article 34, using $h = 2$. Compare results. Show that the latter is exact.

2. Calculate $\int_0^{12} x^7 dx$ by (4), Article 35, using $h = 2$.

   Compare with results in Exercise 1.

3. Calculate $\ln 1.4 = \int_1^{1.4} \frac{dx}{x}$ by (4), Article 34 and by (2), Article 35, using $h = 0.1$. Compare results with the correct value.

4. Calculate $\ln 1.8 = \int_1^{1.8} \frac{dx}{x}$ by (8), Article 34 and by (6), Article 35, using $h = 0.1$. Compare with correct value.

5. Calculate $\int_0^1 \frac{dx}{1+x}$. Select h and a suitable formula to obtain six figure accuracy.

6. Calculate $\int_0^1 \frac{dx}{1+x^2}$ using (5), Article 34, with $h = 0.1$.

7. Use undetermined coefficients to derive (3), Article 35, and determine the error term.

8. Derive (4), Article 34.

9. Show that

$$\int_{x_0}^{x_4} y dx = \frac{4h}{9}(y_0 + 2y_1 + 3y_2 + 2y_2 + y_4)$$

is exact if y is any polynomial of degree three. Investigate the error.

10. Show that

$$\int_{x_0}^{x_6} y dx = 3h(y_1 - y_2 + 2y_3 - y_4 + y_5)$$

is exact if y is any polynomial of degree three. Investigate the error.

Chapter V
NUMERICAL SOLUTION OF DIFFERENTIAL EQUATIONS

As the student of differential equations is well aware, unless a differential equation falls into one of a restricted number of types, it is impossible to express its solution in elementary analytical form. It then becomes necessary to use series or some other approximate method to obtain a particular solution. The aim of this chapter is to present numerical methods by which a solution of a differential equation can be calculated.

The methods explained here are all "step-by-step" methods, so-called because the values of the dependent variable are calculated one after the other for a sequence of equally-spaced values of the independent variable. The successive values of the independent variable x are denoted by $x_0, x_1, x_2, \ldots$, the interval is denoted by h, and the corresponding values of the dependent variable are denoted by $y_0, y_1, y_2, \ldots$. Differentiation with respect to x is indicated by primes, so that a differential equation of the first order, when solved for the derivative, has the form

$$y' = f(x, y).$$

Similarly a differential equation of second order, solved for the second derivative, has the form

$$y'' = f(x, y, y').$$

It is assumed that the function f satisfies all requirements necessary to insure the existence of a unique, continuous, differentiable solution of the form $y =$ function of x throughout the interval under consideration.

37. FIRST METHOD

A very crude but simple method of solving differential equations is based on the second formula on page 96

$$y_{n+1} = y_{n-1} + 2hy'_n.$$

The process is illustrated by an example.

Example. Solve $y' = (1 - y^2)^{\frac{1}{2}}$ with the initial condition $y = 0$, when $x = 0$.

We choose $h = 0.1$, and calculate an approximate value of y_1 using a few terms of Taylor's series. For at $x = 0$,

$$y_0 = 0$$

$$y_0' = (1 - y_0^2)^{\frac{1}{2}} = 1$$

$$y_0'' = -(1 - y_0^2)^{-\frac{1}{2}} y_0 y_0' = 0$$

whence

$$y_1 = 0 + h + 0 = 0.100$$

The computation is arranged according to the following table:

x	y	y'
0	0	1.000
.1	.100	.995
.2		

When the value of y_1 has been found from a few terms of Taylor's series, as shown above, the corresponding value of y' is calculated from the differential equation. In this case $y_1' = .995$.

To proceed we have from (1)

$$y_2 = y_0 + 2hy_1'$$

which gives $y_2 = 0 + 2(.1)(.995) = .199$. Then y_2' is found from the differential equation to be .980. With these entries the table now is

x	y	y'
0	0	1.000
.1	.100	.995
.2	.199	.980
.3		

From (1) $y_3 = y_1 + 2hy_2' = .100 + (.2)(.980) = .296$.

The integration is continued in this manner step by step and, carried to $x = 1.3$, appears in final form as follows:

Numerical Solution of $y' = (1 - y^2)^{\frac{1}{2}}$, $y = 0$ at $x = 0$.

x	y	y'	x	y	y'
0	.000	1.000	.7	.645	.764
.1	.100	.995	.8	.718	.696
.2	.199	.980	.9	.784	.621
.3	.296	.995	1.0	.842	.539
.4	.390	.921	1.1	.892	.452
.5	.480	.877	1.2	.932	.362
.6	.565	.825	1.3	.964	.---

The particular solution of the above equation is $y = \sin x$. Comparing our calculated results with a table of $\sin x$ we find agreement except occasionally for one unit in the third decimal place.

Remarks. 1. The accuracy of the above calculation cannot be improved merely by carrying the work to a greater number of decimal places. The error inherent in the formula employed makes greater accuracy impossible.

2. If greater accuracy is required it can be obtained by shortening the interval h. For example in this particular problem we may obtain accuracy to about six places of decimals by choosing $h = 0.01$, but we do so at the expense of increasing the labor more than ten fold, since ten times as many steps are required, and each step is more laborious.

3. The principal labor in the computation is generally the substitution into the differential equation. Therefore care should be taken at the outset to arrange this substitution in as simple and systematic manner as possible and to make full use of computational aids, such as tables, calculating machines, etc. In the example above a good table of squares and square roots is of great help.

4. The form of the differential equation itself may introduce difficulties which cannot entirely be evaded by refinements in the method of solution. In the example above such a difficulty occurs if y is very near 1, for then a slight error in y produces a large error in the calculated value of y'. For instance if the y used is .998 but should have been .999 we would get $y' = .063$ instead of $y' = .045$, so that a change of .001 in y produces a change of .018 in y'.

5. Evidently any mistake in computation will be carried forward throughout the remainder of the integration. Since there is no check of any sort, it is necessary to guard carefully against mistakes. This is especially true for the calculation of y_1, which is different from the regular routine, and hence more liable to error.

38. SECOND METHOD

The procedure now to be given has two advantages over that of Article 37. First, it permits greater accuracy for the same number of steps than does the other method. Second, it provides checks which indicate the approximate accuracy attainable and also usually catch errors of calculation. This method consists essentially of a step-by-step process using the formula

$$(1) \quad y_{n+1} = y_{n-3} + \frac{4h}{3}(2y_n' - y_{n-1}' + 2y_{n-2}') + \frac{28}{90}h^5y^{(5)}$$

as a "predictor," and Simpson's Rule

$$(2) \quad y_{n+1} = y_{n-1} + \frac{h}{3}(y_{n+1}' + 4y_n' + y_{n-1}') - \frac{1}{90}h^5y^{(5)}$$

as a "corrector." The start of the computation requires four consecutive known values of y. These may be calculated by Taylor's series carried out to terms of the fifth degree in h. However it is usually more satisfactory to find these starting values by successive approximations using the three "starter" formulas

$$(3) \quad y_1 = y_0 + \frac{h}{24}(7y_1' + 16y_0' + y_{-1}') + \frac{h^2y_0''}{4} - \frac{1}{180}h^5y^{(5)},$$

$$(4) \quad y_{-1} = y_0 - \frac{h}{24}(y_1' + 16y_0' + 7y_{-1}') + \frac{h^2y_0''}{4} + \frac{1}{180}h^5y^{(5)},$$

and

$$(5) \quad y_2 = y_0 + \frac{2h}{3}(5y_1' - y_0' - y_{-1}') - 2h^2y_0'' + \frac{7}{45}h^5y^{(5)},$$

together with formula (2) above.

In using these formulas we have x_0 and y_0 given. Then values of y_0' and y_0'' are found from the differential equation and from the equation obtained by differentiation, respectively. Trial values for y_1' and y_{-1}' are given by $y_1' = y_0' + hy_0''$ and $y_{-1}' = y_0' - hy_0''$. From these, with (3) and (4), we obtain trial values of y_1 and y_{-1}, then compute y_1' and y_{-1}' from the differential equation, and recompute y_1 and y_{-1} using the improved values. The process is repeated until no change occurs. Note that $h^2y_0''/4$ is calculated once for all, as it does not change in the recomputations. Note also that the error terms of (3) and (4) are one-half the error term of Simpson's Rule, so that we may be sure that the accuracy of the starting values is fully as good as the subsequent process of integration justifies. Next a trial value of y_2 is calculated by (5), and checked and rechecked by Simpson's rule until no change occurs. The four values y_{-1}, y_0, y_1, y_2, needed for the start of the computation are now ready.

<u>Example</u>. We take the equation $y' = x - y$ with initial values $x_0 = 0$, $y_0 = 2$. Then $y_0' = -2$, $y_0'' = 3$. Using $h = 0.1$ we have the following results:

x	y	y'	
-0.1	2.215	-2.3	
0.0	2.000	-2.0	first approximation
0.1	1.815	-1.7	
-0.1	2.2155	-2.315	
0.0	2.0000	-2.000	second approximation;
0.1	1.8145	-1.715	
-0.1	2.21551	-2.3155	
0.0	2.00000	-2.0000	third approximation
0.1	1.81451	-1.7145	
-0.1	2.21551	-2.31551	
0.0	2.00000	-2.00000	first approximation of y_2
0.1	1.81451	-1.71451	
0.2	1.6562	-1.4562	
-0.1	2.21551	-2.31551	
0.0	2.00000	-2.00000	final values
0.1	1.81451	-1.71451	
0.2	1.65619	-1.45619	

To continue we use (1) to obtain a predicted value of y_3, which proves to be $y_3 = 1.52246$, giving $y_3' = -1.22246$. Using Simpson's Rule (2) we check the value of y_3 obtaining $y_3 = 1.52245$. This is taken as the correct value of y_3 (see Remark 1 below). We correct y_3'.

We next obtain $y_4 = 1.41097$ by (1), calculate y_4', and check with (2), getting $y_4 = 1.41096$, which is taken as correct. In this way we proceed, using (1) to get the trial value of y, then calculating y' from the differential equation, then obtaining the corrected y by (2).

The final computation, as far as $x = 1.2$, is shown below.

$y' = x - y$ $\qquad\qquad$ $y_0 = 2$, when $x = 0$

x	y	y'	D
-.1	2.21551	-2.31551	
0	1.00000	-2.00000	
.1	1.81451	-1.71451	
.2	1.65619	-1.45619	
.3	1.52245	-1.22245	-1
.4	1.41096	-1.01096	-1
.5	1.31959	- .81959	-1
.6	1.24644	- .64644	0
.7	1.18975	- .48975	0
.8	1.14799	- .34799	-1
.9	1.11970	- .21970	-1
1.0	1.10364	- .10364	-1
1.1	1.09860	+ .00140	0
1.2	1.10359	+ .09641	0

<u>Remarks</u>. 1. The difference between the predicted value of y, obtained from (1) and the corrected value obtained by (2) is recorded in the right-hand column. Turning to formulas (1) and (2) we note that the error of (1) is roughly 28 times the error of (2) and in the

opposite direction. So if E_1 is the error of (1), E_2 the error of (2), and $D = E_2 - E_1$ we have $E_2 = D/29$ approximately. Now D is recorded in the right-hand column, and so long as D/29 is not significant we assume that the value given by (2) is correct. Abrupt fluctuations in the values recorded for D indicate the presence of errors and the calculation should be checked.

2. If the error $E_2 = D/29$ proves to be larger than desired accuracy permits, it is necessary to shorten the interval h. Cutting the interval in half will divide the error by about 32.

3. In the example above the correct value of y to six places at $x = 1.2$ is 1.103583, which gives us an idea of the accuracy attained in this particular case.

4. If the entries in the D column are negligible, as in the example above, it is advisable to try an interval twice as great. If sufficient accuracy is attained, the labor can thus be cut in half. To do this we take four consecutive even entries from the result above and proceed with the integration, remembering that h in (1) and (2) is now 0.2. The result is shown below

x	y	y'	D
.6	1.24644		
.8	1.14799	-.34799	
1.0	1.10364	-.10364	
1.2	1.10359	+.09641	
1.4	1.13979	+.26021	-11
1.6	1.20569	+.39431	- 9
1.8	1.29589	+.50411	- 9
2.0	1.40600	+.59400	- 8

The correct value at 2.0 to seven places is $y = 1.4060059$.

39. EQUATIONS OF SECOND ORDER

The extension of the method of the preceding Article

to the case of equations of second order is quite simple. The value of y'_{n+1} is predicted by

$$(1) \qquad y'_{n+1} - y'_{n-3} = \frac{4h}{3}(2y''_n - y''_{n-1} + 2y''_{n-2})$$

and y_{n+1} is then obtained from

$$(2) \qquad y_{n+1} - y_{n-1} = \frac{h}{3}(y'_{n+1} + 4y'_n + y'_{n-1}).$$

With x, y_{n+1}, y'_{n+1} now given we get y'' from the differential equation, then use

$$(3) \qquad y'_{n+1} - y'_{n-1} = \frac{h}{3}(y''_{n+1} + 4y''_n + y''_{n-1})$$

to correct y'_{n+1}. If the correction is significant we must correct y_{n+1} using (2) again, and also correct y''_{n+1}.

The start of the computation is effected by means of the formulas (3), (4), (5), of Article 38, together with the same ones expressed in terms of y', y'', y''' instead of y, y', y''.

Example. Integrate $y'' = -2y'y$ with initial conditions $y = 0$, $y' = 1$, $x = 0$. Here $y''' = -2(y''y + y'^2)$ $= 4y'y^2 - 2y'^2 = -2$ at $x = 0$.

Numerical integration of $y'' = -2yy'$, $y_0 = 0$, $y_0' = 1$.

x	y	y'	y"	D
-.1	-0.099669	.99007	+.19736	
0	.000000	1.00000	.00000	
.1	.099669	.99007	-.19736	
.2	.197377	.96104	-.37937	
.3	.291315	.91514	-.53319	- 9
.4	.379951	.85563	-.65019	-14
.5	.462119	.78645	-.72687	-17
.6	.537051	.71157	-.76430	-17
.7	.604368	.63474	-.76723	-12
.8	.664037	.55905	-.74246	- 8
.9	.716297	.48692	-.69756	- 3
1.0	.761594	.41997	-.63969	0

Remarks. 1. The figure in the sixth place in y, although not reliable, is retained to reduce the danger of accumulated error. These values of y should be rounded off to five places when the computation is completed.

2. The particular solution of the above equation proves to be $y = \tanh x$ and comparing with a table of $\tanh x$ we find that our results above are correct to five places.

40. SPECIAL FORMULAS FOR SECOND ORDER EQUATIONS

Many differential equations of the second order occurring in physics, mechanics, and astronomy are of the type

$$y'' = f(x, y),$$

which lacks the first derivative. Special methods are available which shorten the labor in solving equations of this type.

The first is a crude but rapid method, entirely similar to that given in Article 37, and based on the formula

$$y_{n+1} = 2y_n - y_{n-1} + h^2 y_n'' + \frac{1}{12}h^4 y^{(4)}$$

The procedure is so nearly similar to the one in Article 37, that we shall not elaborate further.

The second resembles the method of Article 38. It uses the formula

$$(1) \qquad y_{n+1} = y_n + y_{n-2} - y_{n-3} + \frac{h^2}{4}(5y_n'' + 2y_{n-1}'' + 5y_{n-2}'') + \frac{17}{240}h^6 y^{(6)}$$

as a predictor and

$$(2) \qquad y_n = 2y_{n-1} - y_{n-2} + \frac{h^2}{12}(y_n'' + 10y_{n-1}'' + y_{n-2}'') - \frac{1}{240}h^6 y^{(6)}$$

as a corrector. Four consecutive starting values are required, which may be obtained as in Articles 38 and 39. As an example we give a few steps in the solution of $y'' = -y^3$ with initial values $x = 0$, $y = 0$, $y' = 1$.

x	y	y"	D
0.2	0.1999840	-0.007998	
0.3	0.2998785	-0.026967	
0.4	0.3994885	-0.063755	1
0.5	0.4984415	-0.123835	2
0.6	0.5961329	-0.211850	7
0.7	0.6916799	-0.330914	12

Remarks. 1. In this case the starting values were found from Taylor's Series; the first three terms of which are

$$y = x - \frac{1}{20}x^5 + \frac{1}{480}x^9 + \ldots$$

2. The column headed D plays the same role here as in Article 38, except that the error E_2 of (2) is approximately D/18.

41. SIMULTANEOUS EQUATIONS

The extension of the foregoing methods to the case of simultaneous equations is almost obvious. Each step in the process of integration is carried out independently for each equation just as though it were a single equation except for the substitution into the equations. For example, in the case of a pair of simultaneous equations of the second order

$$x'' = f(x, y, x', y', t)$$
$$y'' = g(x, y, x', y', t),$$

in which accents denote differentiation with respect to the independent variable t, we integrate the first equation to find x' and x, the second to find y' and y, and substitute the four values obtained to find x" and y".

The following example illustrates the process.

Example. Solve the equations

$$x'' = -2x + y$$
$$y'' = -2y + x$$

with the conditions $x = 1$, $x' = 0$, $y = 0$, $y' = 1$ at $t = 0$.

We choose $h = 0.1$, and for simplicity use the crude formulas

$$x_{n+1} = 2x_n - x_{n-1} + h^2 x_n'',$$
$$y_{n+1} = 2y_n - y_{n-1} + h^2 y_n''.$$

To get started we calculate x_1 and y_1 from a few terms of the Taylor's series. The computation appears below.

t	x	x"	y	y"
0	1.0000	-2.000	.0000	1.000
.1	.9902	-1.876	.1046	.781
.2	.9616	-1.706	.2161	.527
.3	.9160	-1.497	.3348	.246
.4	.8554	-1.256	.4550	- .055
.5	.7822	- .990	.5747	- .367
.6	.6992	- .708	.6906	- .682
.7	.6090	- .418	.7998	- .991
.8	.5147	- .130	.8990	-1.283
.9	.4191	+ .147	.9854	-1.552
1.0	.3249		1.0563	

Remark. The figures in the fourth decimal place are not reliable, and after the calculation x and y should be rounded off to three decimal places.

42. USE OF FIVE-TERM FORMULAS

When the labor of performing the substitutions into the differential equation is considerable, it is sometimes advantageous to use more accurate formulas for the integration rather than to shorten the interval in order to secure the desired accuracy in the final result. Thus the method of Article 38 may be modified by using

$$(1) \qquad y_{n+1} = y_{n-5} + \frac{3h}{10}(11y'_n - 14'_{n-1} + 26y'_{n-2} - 14'_{n-3} + 11y'_{n-4})$$

as a predictor and

$$(2) \qquad y_{n+1} = y_{n-3} + \frac{2h}{45}(7y'_{n+1} + 32y'_n + 12y'_{n-1} + 32y'_{n-2} + 7y'_{n-3})$$

as a corrector. The error of (1) is $\frac{41}{140} h^7y^{(7)}$ and that of (2) is $\frac{-8}{945} h^7y^{(7)}$, so that the error of (2) is approximately D/28, where D has the same significance as in Article 38. We can thus estimate the degree of accuracy attainable. It is usually best to start the computation with a simpler method, say that of Article 38, using suitable intervals, and when enough values have been obtained double the interval and use (1) and (2) of this article.

Similarly the method of Article 40 may be modified by using

$$(3) \qquad y_{n+1} = y_n + y_{n-4} - y_{n-5} + \frac{h^2}{48}(67y''_n - 8y''_{n-1} + 122y''_{n-2} - 8y''_{n-3} + 67y''_{n-4})$$

as a predictor and

$$(4) \qquad y_{n+1} = y_n + y_{n-2} - y_{n-3} + \frac{h^2}{240}(17y''_{n+1} + 232y''_n + 222y''_{n-1} + 232y''_{n-2} + 17y''_{n-3})$$

as a corrector. These formulas are exact if y is any polynomial of the seventh degree.

Exercises

Using the crude method of Article 37, solve to three decimal places:

1. $y' = 1 - xy$, $y = 0$ at $x = 0$, $h = 0.1$ from $x = 0$ to $x =$
2. $y' = x^2 - y^2$, $y = 0$ at $x = 1$, $h = 0.1$, from $x = 1$ to $x = 2$.
3. $y' = \sqrt{1 + xy}$, $y = 0$ at $x = 0$, $h = 0.05$ from $x = 0$ to $x = 1$. Using the method of Article 38, solve to five decimal places:
4. $y' = 1 - y^2$, $y = 0$ at $x = 0$, $h = 0.1$, from $x = 0$ to $x = 1$.
5. $y' = \sqrt{1 + y^3}$, $y = 0$ at $x = 0$, $h = 0.1$, from $x = 0$ to $x = 1$.
6. $y' = \log_{10} xy$, $y = 1$ at $x = 1$, $h = 0.2$ from $x = 1$ to $x = 2$.

   Integrate the following:
7. $y'' - \frac{1}{2}y'^2 + y = 0$ from $y = 1$, $y' = 0$, to the point where y' is again zero.
8. $y'' = \sin y$ from $y = 50^\circ$, $y' = 0$ to the point where $y = 0$.
9. $x'' = x'y$, $y'' = y'x$, $x = y = 1$, $x' = y' = 1$ at $t = 0$.

Chapter VI
FINITE DIFFERENCES*

The calculus of finite differences has many analogies to the infinitesimal calculus, and is much used in numerical calculations. Interpolation, subtabulation detection of errors, and numerical integration are examples of numerical processes in which the method of finite differences plays an important role. Before going into the subject of differences, we first consider factorial polynomials and binomial coefficients.

43. FACTORIAL POLYNOMIALS

A polynomial of degree n of the form

$$x(x - 1)(x - 2) \ldots (x - n + 1)$$

is called a factorial polynomial and is represented by the notation $x^{(n)}$, so that

$$(1) \quad x^{(n)} = x(x - 1)(x - 2) \ldots (x - n + 1)$$

In particular, we set $x^{(0)} = 1$. Evidently $x^{(n)} = 0$ for $x = 0, 1, 2, \ldots, n - 1$, while if x is an integer greater than $n - 1$

$$x^{(n)} = \frac{x!}{(x - n)!}$$

If the product on the right in (1) is multiplied out the result may be written as a polynomial expressed in terms of descending powers of x.

$$(2) \quad x^{(n)} = S_0^n x^n + S_1^n x^{n-1} + \ldots + S_{n-1}^n x,$$

*See Appendix C.

in which the coefficients S_i^n are known as <u>Stirling's Numbers of the first kind</u>. Because of the relation

$$x^{(n+1)} = x^{(n)}(x - n)$$

it follows from (2) that

$$S_i^{n+1} = S_i^n - nS_{i-1}^n \tag{3}$$

This recurrence formula affords a simple means of building up a table of Stirling's numbers, as follows:

n	S_0^n	S_1^n	S_2^n	S_3^n	S_4^n	S_5^n	S_6^n	S_7^n	S_8^n	S_9^n
0	1	0	0							
1	1	0	0							
2	1	-1	0							
3	1	-3	2							
4	1	-6	11	-6						
5	1	-10	35	-50	24					
6	1	-15	85	-225	274	-120				
7	1	-21	175	-735	1624	-1764	720			
8	1	-28	322	-1960	6769	-13132	13068	-5040		
9	1	-36	546	-4536	22449	-67284	118124	-109584	40320	
10	1	-45	870	-9450	63273	-269325	723680	-1172700	1026576	-3628800

Any entry in the table is equal to the number directly above minus the number above and to the left multiplied by the n in that row.

For example, 6769 = 1624 - (7)(-735). Having computed such a table, we may at once write down any factorial in the form (2). For example

$$x^{(7)} = x^7 - 21x^6 + 175x^5 - 735x^4 + 1624x^3 - 1764x^2 + 720x.$$

Any polynomial expressed in terms of powers of x may equally well be expressed in terms of factorials.

<u>Example 1</u>. Express the polynomial

$$P(x) = 3x^5 - 7x^4 + 87x^3 + 28x^2 + 176x - 77$$

in terms of factorials.

Evidently the factorial of highest degree will be $x^{(5)}$ so we subtract $3x^{(5)}$ from the given polynomial leaving a polynomial of degree 4 whose leading term is $23x^4$. From this we subtract $23x^{(4)}$ leaving a polynomial of degree 3 with leading term $120x^3$. Proceeding step-by-step in this manner, we arrive finally at a zero remainder and the original polynomial must therefore be equal to the sum of the terms subtracted. The work, systematically arranged, is shown below:

$P(x) =$	3	- 7	+ 87	+ 28	+ 176	- 77
$3x^{(5)} =$	3	- 30	+ 105	- 150	72	
Diff. =		23	- 18	+ 178	+ 104	- 77
$23x^{(4)} =$		23	- 138	+ 253	- 138	
Diff. =			120	- 75	+ 242	- 77
$120x^{(3)} =$			120	- 360	+ 240	
Diff. =				285	+ 2	- 77
$285x^{(2)} =$				285	- 285	
Diff. =					287	- 77
$287x^{(1)} =$					287	
Diff. =						- 77
$-77x^{(0)} =$						- 77

Hence, the desired form of the polynomial is

$$P(x) = 3x^{(5)} + 23x^{(4)} + 120x^{(3)} + 285x^{(2)} + 287x^{(1)} - 77x^{(0)}.$$

While the foregoing method is relatively simple the same result may be obtained more elegantly by means of differences as explained in Article 47.

Factorial polynomials satisfy a number of useful identities, one of which is

$$(x + 1)^{(n)} - x^{(n)} = nx^{(n-1)} \tag{4}$$

For the left-hand member may be written

$$\begin{aligned} &(x + 1)x(x - 1) \ldots (x - n + 2) \\ &\qquad -x(x - 1) \ldots (x - n + 2)(x - n + 1) \\ &= \Big[(x + 1) - (x - n + 1)\Big] x(x - 1) \ldots (x - n + 2) \\ &= nx^{(n-1)} \end{aligned}$$

Another formula is

$$\sum_{j=0}^{k}(x + j)^{(n)} = \frac{(x + k + 1)^{(n+1)} - (x)^{(n+1)}}{(n + 1)} \tag{5}$$

This may be derived from (4) by first replacing n with $(n + 1)$ and then replacing x successively by x, $x + 1$, $x + 2, \ldots, x + k$ so as to obtain the equations

$$\begin{aligned} (x + 1)^{(n+1)} - x^{(n+1)} &= (n + 1)x^{(n)} \\ (x + 2)^{(n+1)} - (x + 1)^{(n+1)} &= (n + 1)(x + 1)^{(n)} \\ (x + 3)^{(n+1)} - (x + 2)^{(n+1)} &= (n + 1)(x + 2)^{(n)} \\ \text{-----------} & \text{-----------} \\ (x + k + 1)^{(n+1)} - (x + k)^{(n+1)} &= (n + 1)(x + k)^{(n)}. \end{aligned}$$

When we add these and divide by $(n + 1)$ we obtain (5).

If $P(x)$ is a polynomial we may determine the value of a sum

$$\sum_{j=m}^{j=n} P(j)$$

by means of equation (5).

<u>Example 2</u>.

Find $$\sum_{j=0}^{k} j^3.$$

First, we express the function x^3 in terms of factorial polynomials and by the method of example 1 obtain

$$x^3 = x^{(3)} + 3x^{(2)} + x^{(1)}.$$

Then

$$\sum_{j=0}^{k} (x + j)^3 = \sum_{j=0}^{k} (x + j)^{(3)}$$

$$+ 3\sum_{j=0}^{k} (x + j)^{(2)} + \sum_{j=0}^{k} (x + j)^{(1)}.$$

Applying formula (5) to the sums on the right we get

$$\sum_{j=0}^{k} (x + j)^{(3)} = \frac{(x + k + 1)^{(4)} - x^{(4)}}{4}$$

$$+ 3\ \frac{(x + k + 1)^{(3)} - x^{(3)}}{3} + \frac{(x + k + 1)^{(2)} - x^{(2)}}{2}$$

and finally, setting $x = 0$, we have

$$\sum_{j=0}^{k} j^3 = \frac{(k + 1)^{(4)}}{4} + (k + 1)^{(3)} + \frac{(k + 1)^{(2)}}{2}$$

$$= \frac{1}{4}(k + 1)k\left[(k - 1)(k - 2) + 4(k - 1) + 2\right]$$

$$= \frac{(k + 1)^2 k^2}{4}.$$

Exercises

1. Show that

$$(x + 2)^{(n)} - 2(x + 1)^{(n)} + x^{(n)} = n(n - 1)x^{(n-2)}.$$

2. Find the value of $\sum_{j=0}^{10} j^2$

3. Find the value of $\sum_{j=0}^{10} (j - 3)^2$

4. Find the value of $\sum_{j=0}^{7} (j + \frac{1}{2})^2$

5. Express in terms of factorial polynomials

$$\sum_{1}^{k} (j^3 + j^2 + j)$$

44. THE BINOMIAL COEFFICIENT FUNCTIONS

The binomial theorem states that

$$(1) \qquad (a + b)^s = a^s + sa^{s-1} b + \frac{s(s - 1)}{2!}a^{s-2}b^2 + \frac{s(s - 1)(s - 2)}{3!}a^{s-3} b^3 + \frac{s(s - 1)(s - 2)(s - 3)}{4!}a^{s-4} b^4 + \dots$$

If s is a positive integer or zero the series on the right in (1) terminates with the $(s + 1)$th term, which is b^s. For all other real values of s the series continues without end. The coefficients in the series have many interesting properties and occur in a great number of situations. The standard notation used here is

$$\binom{s}{k} = \frac{s(s - 1) \dots (s - k + 1)}{k!} \qquad (k \text{ an integer}),$$

so that in particular

$$\binom{s}{0} = 1, \quad \binom{s}{1} = s, \quad \binom{s}{2} = \frac{s(s - 1)}{2!}, \quad \text{etc.}$$

We are particularly interested in these expressions as functions of the variable s, and observe that the binomial coefficient function $\binom{s}{k}$ is a polynomial in s of degree k which vanishes at the k values, s = 0, 1, 2, . . ., k - 1. In fact, if we use the notation for factorial polynomials, it is clear that

$$\binom{s}{k} = \frac{s^{(k)}}{k!}.$$

For brevity we shall often refer to the binomial coefficient functions simply as 'binomial coefficients.'

Using the notation $\binom{s}{k}$ for the coefficients, we may write (1) in the form

$$(2) \qquad (a + b)^s = \sum_{k=0}^{\infty} \binom{s}{k} a^{s-k} b^k.$$

Similarly

$$(3) \qquad (1 + x)^s = \sum_{k=0}^{\infty} \binom{s}{k} x^k.$$

This series is absolutely convergent for $|x| < 1$.

The binomial coefficients satisfy an important identity,

$$(4) \qquad \binom{s+t}{k} = \sum_{j=0}^{k} \binom{s}{j} \binom{t}{k-j} = \sum_{j=0}^{k} \binom{s}{k-j} \binom{t}{j}.$$

This relationship is proved by multiplying the two absolutely convergent series

$$(1 + x)^s = \sum_{n=0}^{\infty} \binom{s}{n} x^n, \qquad |x| < 1,$$

$$(1 + x)^t = \sum_{m=0}^{\infty} \binom{t}{m} x^m, \qquad |x| < 1.$$

The product of the quantities on the left is $(1 + x)^{s+t}$, and after the product of the two series on the right is arranged in ascending powers of x, the coefficient of x^k proves to be

$$\sum_{j=0}^{k} \binom{s}{j} \binom{t}{k-j} .$$

On the other hand

$$(1 + x)^{s+t} = \sum_{k=0}^{\infty} \binom{s+t}{k} x^k,$$

so that the coefficient of x^k is $\binom{s+t}{k}$. Since these coefficients must be identical we have

$$\binom{s+t}{k} = \sum_{j=0}^{k} \binom{s}{j} \binom{t}{k-j} .$$

If $t = 1$ in equation (4) we have

$$(5) \qquad \binom{s+1}{k} = \binom{s}{k} + \binom{s}{k-1} .$$

If n is an integer we see that

$$(6) \qquad \binom{n}{k} = 0 \qquad \text{if } k > n.$$

It is also convenient to define $\binom{s}{k}$ as zero for all negative integral values of k, so that

$$(7) \qquad \binom{s}{k} = 0 \qquad \text{if } k < 0.$$

For the integral values of s relation (5), together with (6) and (7), enables us to build up by simple additions a table of the binomial coefficients in the familiar triangular scheme where each entry is the sum of the two adjacent entries in the row above:

1

1 1

1 2 1

1 3 3 1

1 4 6 4 1

1 5 10 10 5 1

1 6 15 20 15 6 1

<u>Exercises</u>

Show that

1. If n is an integer $\binom{n}{k} = \binom{n}{n-k}$

2. $$\binom{s+1}{k+1} = \frac{s+1}{k+1}\binom{s}{k}$$

3. $$\binom{s+1}{k} = \frac{s+1}{s-k+1}\binom{s}{k}$$

4. $$\binom{s}{k+1} = \frac{s-k}{k+1}\binom{s}{k}$$

5. $$\binom{s+1}{k} - \binom{s}{k} = \binom{s}{k-1},$$

$$\binom{s+2}{k} - 2\binom{s+1}{k} + \binom{s}{k} = \binom{s}{k-2},$$

$$\binom{s+3}{k} - 3\binom{s+2}{k} + 3\binom{s+1}{k} - \binom{s}{k} = \binom{s}{k-3},$$

and in general

$$\sum_{j=0}^{q} (-1)^j \binom{q}{j}\binom{s+q-j}{k} = \binom{s}{k-q} \quad \text{(q an integer)}$$

6. $$\sum_{k=0}^{n} \binom{n}{k} = 2^n \qquad \text{(n an integer)}$$

$$\sum_{k=0}^{n} (-1)^k \binom{n}{k} = 0 \qquad \text{(n an integer)}$$

7. $$\sum_{k=0}^{n} \binom{n}{k}^2 = \binom{2n}{n} \qquad \text{(n an integer)}$$

8. $$\binom{-s}{k} = (-1)^k \binom{s+k-1}{k}$$

9. $$\binom{s-t}{k} = \sum_{j=0}^{k} (-1)^j \binom{s}{k-j} \binom{t+j-1}{j}$$

10. $$\binom{s}{k} \binom{s+q}{q} = \binom{k+q}{k} \binom{s+q}{k+q}$$

45. FINITE DIFFERENCES

Let $y_0, y_1, y_2, \ldots$ be the values of a function $y = f(x)$ corresponding to equally spaced values $x_0, x_1, x_2, \ldots$ of the independent variable x. The following notation* will be used to designate differences:

$$y_1 - y_0 = \Delta y_0$$

$$y_2 - y_1 = \Delta y_1$$

- - - - - - - - - -

$$y_n - y_{n-1} = \Delta y_{n-1}.$$

The subscript attached to the Δy is taken the same as the subscript of the <u>second</u> member of the difference.

The differences designated above are called forward differences of the first order or simply <u>first differences</u>. Second order differences are denoted as follows:

*See Appendix A.

$$\Delta y_1 - \Delta y_0 = \Delta^2 y_0$$

$$\Delta y_2 - \Delta y_1 = \Delta^2 y_1$$

etc.

and in general $(n+1)^{th}$ order differences are obtained from those of nth order by the formulas

$$\Delta^n y_1 - \Delta^n y_0 = \Delta^{n+1} y_0$$

$$\Delta^n y_2 - \Delta^n y_1 = \Delta^{n+1} y_1$$

etc.

By successive substitutions we easily show that

$$\Delta^2 y_0 = y_2 - 2y_1 + y_0$$

$$\Delta^3 y_0 = y_3 - 3y_2 + 3y_1 - y_0$$

and in general

(1) $$\Delta^n y_0 = \sum_{k=0}^{n} (-1)^k \binom{n}{k} y_{n-k}.$$

In a similar manner, by successive eliminations from equations of type (1), we obtain

(2) $$y_n = y_0 + n\Delta y_0 + \frac{n(n-1)}{2!} \Delta^2 y_0 + \cdots + \Delta^n y_0$$

$$= \sum_{k=0}^{n} \binom{n}{k} \Delta^k y_0.$$

Formula (2) may be expressed in the following easily remembered symbolic form

(3) $$y_n = (1 + \Delta)^n y_0$$

A difference table is set up according to the scheme below.

x_0	y_0				
		Δy_0			
x_1	y_1		$\Delta^2 y_0$		
		Δy_1		$\Delta^3 y_0$	
x_2	y_2		$\Delta^2 y_1$		$\Delta^4 y_0$
		Δy_2		$\Delta^3 y_1$	
x_3	y_3		$\Delta^2 y_2$		___
		Δy_3		___	
x_4	y_4		___		
		___			
___	___				

Each entry in the table of differences is the difference of the adjacent entries in the column to the left.

<u>Example</u>. Form a table of differences for $y = \sin x$ with the entries $x = 0, 0.1, 0.2, 0.3, 0.4, 0.5, 0.6$.

The table appears as follows:

x	y				
0	.00000				
		9983			
0.1	.09983		-99		
		9884		-100	
0.2	.19867		-199		4
		9685		-96	
0.3	.29552		-295		2
		9390		-94	
0.4	.38942		-389		3
		9001		-91	
0.5	.47943		-480		
		8521			
0.6	.56464				

Note that in the columns of differences it saves space to omit the decimal point, and this omission need cause no confusion since we can easily restore the decimal point if needed. It is also customary not to carry the formation of higher differences beyond the point where they begin to show signs of fluctuating irregularly. Thus we would not ordinarily use differences beyond the fourth in the example just shown.

If $y = P(x)$ is any polynomial of degree n in x whose values $y_0, y_1, y_2, \ldots$, are tabulated for equally spaced values $x_0, x_1, x_2, \ldots$, with interval h, we may make the change of variable $x = hs + x_0$, after which P(x) becomes a polynomial Q(s) of degree n in s such that

$$y_k = Q(k), \qquad k = 0, 1, 2, \ldots$$

The polynomial Q(s) can be expressed in terms of factorials

$$Q(s) = a_0 s^{(n)} + a_1 s^{(n-1)} + \ldots + a_n.$$

Now formula (4) of Article 43 gives

$$\Delta s^{(n)} = ns^{(n-1)}$$

and in general

$$\Delta^k s^{(n)} = n^{(k)} s^{(n-k)}.$$

Hence

$$\Delta_h{}^k \, P(x) = \Delta_1{}^k \, Q(s) = a_0 n^{(k)} s^{(n-k)}$$

$$+ a_1 (n-1)^{(k)} s^{(n-1-k)} + \ldots,$$

where the subscript on the Δ indicates the intervals for which the differences are formed. It is evident that the k-th difference of any polynomial of degree n is a polynomial of degree n - k. In particular the nth difference is a constant and all differences of order higher than n are zero.

If f(x) is any function, not necessarily a polynomial, with continuous derivative of order n the operation $\Delta^n f(x)$ will be seen to satisfy the conditions for the operator R(f) treated in Article 30. Its value may be determined as in Article 31 and the result is

$$\Delta^n f(x) = f^{(n)}(z)h^n,$$

in which z lies between the greatest and least of the values used in forming $\Delta^n f(x)$. For any given function therefore the nth difference varies roughly as the nth power of the interval. For a given interval the nth difference varies roughly as the n-th derivative.

46. DETECTION OF ERRORS

The method of differences is frequently employed to check a table of functional values. If such a table is to be useful, the tabular interval h must be chosen with respect to the number of decimal places in such a way that the nth differences are negligible to the given number of decimal places for a comparatively low value of n. Suppose that a table has been calculated, the differences formed, and suppose that differences of say the fifth order are negligible. If some entry in the table has an error of magnitude E, the effect of this error will propagate itself through the difference table in the manner shown below:

y	Δy	$\Delta^2 y$	$\Delta^3 y$	$\Delta^4 y$	$\Delta^5 y$
—		—		—	
	—		—		— E
—		—		— E	
	—		— E		— -5E
—		— E		— -4E	
	— E		— -3E		— 10E
— E		— -2E		— 6E	
	— -E		— 3E		— -10E
—		— E		— -4E	

y	Δy	$\Delta^2 y$	$\Delta^3 y$	$\Delta^4 y$	$\Delta^5 y$
	——		—— $-E$		—— $5E$
——		——		—— E	
	——		——		—— $-E$
——		——		——	

Since fifth differences are normally negligible, the effect of the error on the fifth differences becomes quite conspicuous, and we may not only locate the entry that is incorrect but even estimate the amount of the error. Of course, if many entries have errors, this becomes complicated.

In the use of finite differences there is another source of irregularity which is not due to actual errors but to the inaccuracies caused by rounding off the entries to a given number of decimal places. The maximum possible effect due to this cause would occur if quantities neglected in each entry alternated in sign and were equal to half a unit in the last decimal place retained. If N is the number of decimals retained, we set $e = \frac{1}{2}10^{-N}$ and consider the difference table

y	Δy	$\Delta^2 y$	$\Delta^3 y$	$\Delta^4 y$
—— $+e$				
	—— $-2e$			
—— $-e$		—— $+4e$		
	—— $+2e$		—— $-8e$	
—— $+e$		—— $-4e$		—— $+16e$
	—— $-2e$		—— $+8e$	
—— $-e$		—— $+4e$		
	—— $+2e$			
—— $+e$				

It appears that the maximum possible effect on the n-th difference due to rounding off is $2^n e$, or $2^{n-1}10^{-N}$.

Because of fluctuations due to neglecting digits beyond the Nth decimal place the higher ordered differences will not actually approach zero but will begin to fluctuate irregularly and to increase in magnitude as the order of differences increases. There is obviously no sense in calculating a table of differences beyond the point where these fluctuations begin to mask the actual trend of the differences. In this connection the expression $2^{n-1}10^{-N}$ obtained above gives some clue as to whether irregularities are due to actual errors in the tabulated values or merely to dropping decimals beyond the Nth.

Exercises

Locate and correct the error in the following table:

x	f(x)	x	f(x)
0.5	0.19146	1.1	0.36433
0.6	0.22575	1.2	0.38493
0.7	0.25804	1.3	0.40320
0.8	0.28814	1.4	0.41924
0.9	0.31594	1.5	0.43319
1.0	0.34124	1.6	0.44520

47. NEWTON'S BINOMIAL INTERPOLATION FORMULA, FORWARD DIFFERENCES

An interesting and useful set of interpolation formulas can be derived by supposing that the interpolating polynomial is expressed in terms of factorial polynomials. To do this we first make the change of variable $x = hs + x_0$, where h is the interval between the equally spaced values of x, one of which is x_0. Let y_s be the interpolating polynomial of degree n, and let

$$y_s = a_0 + a_1 s + a_2 s^{(2)} + \ldots + a_n s^{(n)} \tag{1}$$

where the a's are undetermined coefficients. Then if we apply the operator Δ^k to equation (1) we have

(2) $$\Delta^k y_s = \sum_{i=0}^{n} a_i \quad \Delta^k s^{(i)}.$$

In this equation let $s = 0$. Now

$$\Delta^k s^{(i)} \equiv 0 \quad \text{if} \quad i < k$$

$$\Delta^k s^{(k)} = k!$$

$$\Delta^k s^{(i)} = 0 \quad \text{for } s = 0 \quad \text{if} \quad i > k.$$

Hence (2) becomes

$$\Delta^k y_0 = k!a_k$$

and

$$a_k = \Delta^k y_0/k!$$

With these values for the coefficients (1) becomes

(3) $$y_s = y_0 + \Delta y_0 s + \Delta^2 y_0 \binom{s}{2} + \cdots + \Delta^n y_0 \binom{s}{n}.$$

This is Newton's Binomial Interpolation Formula. It can be expressed in a neat symbolic form as the first $n + 1$ terms in the expansion of

$$(1 + \Delta)^s y_0.$$

Note that the values of y actually entering into the formation of (3) are $y_0, y_1, \ldots, y_n$.

48. NEWTON'S BINOMIAL INTERPOLATION FORMULA, BACKWARD DIFFERENCES

Here we assume that the interpolating polynomial is of the form

(1) $$y_s = a_0 + a_1 s + a_2(s + 1)^{(2)} + a_3(s + 2)^{(3)} + \cdots + a_n(s + n - 1)^{(n)}$$

Then

$$\Delta^k y_s = \sum_{i=0}^{n} a_i \quad \Delta^k (s + i - 1)^{(i)}$$

and again

$$\Delta^k (s + i - 1)^{(i)} \equiv 0 \qquad \text{if} \quad i < k$$

$$\Delta^k (s + k - 1)^{(k)} = k!$$

$$\Delta^k (s + i - 1)^{(i)} = 0 \qquad \text{for} \quad s = -k \quad \text{if} \quad i > k$$

so that when $s = -k$ we have

$$\Delta^k y_{-k} = k!\ a_k,$$

and equation (1) becomes

$$(2) \qquad y_s = y_0 + \Delta y_{-1} s + \Delta^2 y_{-2} \binom{s+1}{2} + \Delta^3 y_{-3} \binom{s+2}{3} + \ldots + \Delta^n y_{-n} \binom{s+n-1}{n}.$$

The values of y which enter this formula are y_0, y_{-1}, y_{-2}, . . . , y_{-n}. Since $\binom{s+k-1}{k} = (-1)^k \binom{-s}{k}$ the substitution $t = -s$ enables us to write (2) in the form

$$(3) \qquad y_s = y_0 - \Delta y_{-1} t + \Delta^2 y_{-2} \binom{t}{2} - \ldots + (-1)^n \Delta^n y_{-n} \binom{t}{n},$$

where $s = -t = (x - x_0)/h$.

This is Newton's Binomial Interpolation Formula with Backward Differences. Since the starting point for numbering the subscripts is immaterial, (3) may equally well be written

$$(4) \qquad y_s = y_n - \Delta y_{n-1} t + \Delta^2 y_{n-2} \binom{t}{2} - \cdots + (-1)^n \Delta^n y_0 \binom{t}{n},$$

where $s = -t = (x - x_n)/h$, and where the values of y involved are $y_0, y_1, \ldots, y_n$.

49. GAUSS'S INTERPOLATION FORMULAS

The device used to derive Newton's formulas can be extended to secure a variety of formulas where the interpolating polynomial is expressed in terms of binomial coefficient functions.

Let us suppose that the interpolating polynomial is expressed in the form

$$(1) \qquad y_s = a_0 + a_1 \binom{s + m_0}{1} + a_2 \binom{s + m_1}{2} + a_3 \binom{s + m_2}{3} + \cdots$$

in which the m's are integers chosen in such a way that each binomial coefficient function contains all the linear factors occurring in the preceding coefficient. An examination of the factors will reveal that in order to satisfy this condition we must have either $m_{i+1} = m_i$, or $m_{i+1} = m_i + 1$. Applying the operator Δ^k to (1) we get

$$(2) \qquad \Delta^k y_s = a_k + a_{k+1} \binom{s + m_k}{1} + a_{k+2} \binom{s + m_{k+1}}{2} + \cdots$$

If now we let $s = -m_k$, and remember that by our choice of the m's each coefficient will contain the factor $s + m_k$, we see that (2) reduces to

$$\Delta^k y_{-m_k} = a_k.$$

With these values of the a's equation (1) becomes

$$(3) \qquad y_s = y_{-m_0} + \Delta y_{-m_1}\binom{s+m_0}{1} + \Delta^2 y_{-m_2}\binom{s+m_1}{2} + \Delta^3 y_{-m_3}\binom{s+m_2}{3} + \dots$$

As an illustration of the use of equation (3) let us start with $m_0 = 0$. Then we have two choices for m_1, either 0 or 1. Take $m_1 = 1$. Then m_2 is either 1 or 2. Take $m_2 = 1$. Continue in this manner and suppose that the sequence of m's chosen is 0, 1, 1, 2, 2, 2. Then (3) becomes

$$y_s = y_0 + \Delta y_{-1}\binom{s}{1} + \Delta^2 y_{-1}\binom{s+1}{2} + \Delta^3 y_{-2}\binom{s+1}{3} + \Delta^4 y_{-2}\binom{s+2}{4} + \Delta^5 y_{-2}\binom{s+2}{5}.$$

Note that the differences used in any particular choice of (3) are found from the difference table by proceeding toward the right through the table with diagonal moves as in checkers. Note also that the binomial coefficient function in any term is completely determined by the difference occurring in the preceding term. For example if a term contains $\Delta^3 y_{-2}$, which depends on y_{-2}, y_{-1}, y_0, y_1, the coefficient in the next term will be

$$\frac{(s+2)(s+1)\,s\,(s-1)}{4!}.$$

The relation between the subscripts of the y's and the factors of the coefficient in the next term is apparent.

Equation (3) and the rules for the formation of the m's constitute in effect what are known as <u>Sheppard's Rules</u>. By means of these rules we may readily construct a whole set of interpolating polynomials expressed in various forms in terms of differences.

In Newton's forward formula all m's are 0, and in the backward formula $m_i = i$, $i = 0, 1, 2, \ldots, n$. These are the two extreme choices, while all other possibilities are intermediate cases. Two of these will be considered: a) Let the m's be 0, 0, 1, 1, 2, 2, 3, 3, etc. The corresponding form of (1) is

$$(4) \quad y_s = y_0 + \Delta y_0 s + \Delta^2 y_{-1}\binom{s}{2} + \Delta^3 y_{-1}\binom{s+1}{3} + \Delta^4 y_{-2}\binom{s+1}{4} + \Delta^5 y_{-2}\binom{s+2}{5} + \cdots$$

This is Gauss's Forward Formula. b) Let the m's be 0, 1, 1, 2, 2, 3, 3, 4, etc. Then we have Gauss's Backward Formula

$$(5) \quad y_s = y_0 + \Delta y_{-1} s + \Delta^2 y_{-1}\binom{s+1}{2} + \Delta^3 y_{-2}\binom{s+1}{3} + \Delta^4 y_{-2}\binom{s+2}{4} + \Delta^5 y_{-3}\binom{s+2}{5} + \cdots$$

To make clear the relationship between the four formulas, Newton's forward and backward and Gauss's forward and backward, let us set each one up for the same set of values $y_0, y_1, \ldots, y_6$. First of all, the table of differences is calculated as follows:

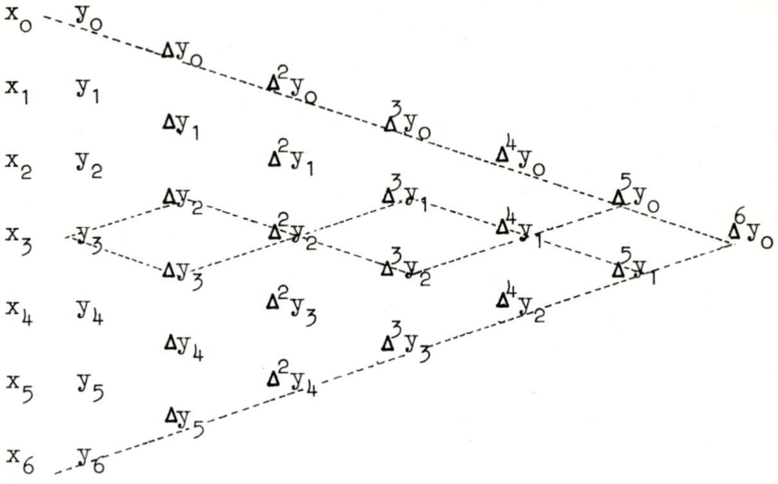

Then Newton's forward difference interpolation formula is

$$y_s = y_0 + \Delta y_0 \binom{s}{1} + \Delta^2 y_0 \binom{s}{2} + \Delta^3 y_0 \binom{s}{3}$$
$$+ \Delta^4 y_0 \binom{s}{4} + \Delta^5 y_0 \binom{s}{5} + \Delta^6 y_0 \binom{s}{6}$$

in which $s = \underline{(x - x_0)}/h$. This uses the differences in the top diagonal.

Newton's backward difference interpolation formula is

$$y_s = y_6 - \Delta y_5 \binom{s}{1} + \Delta^2 y_4 \binom{s}{2} - \Delta^3 y_3 \binom{s}{3}$$
$$+ \Delta^4 y_2 \binom{s}{4} - \Delta^5 y_1 \binom{s}{5} + \Delta^6 y_0 \binom{s}{6},$$

in which $s = (x_6 - x)/h$. This uses the differences in the bottom diagonal.

Gauss's forward formula is

$$y_s = y_3 + \Delta y_3 \binom{s}{1} + \Delta^2 y_2 \binom{s}{2} + \Delta^3 y_2 \binom{s+1}{3} + \Delta^4 y_1 \binom{s+1}{4} + \Delta^5 y_1 \binom{s+2}{5} + \Delta^6 y_0 \binom{s+2}{6}$$

and Gauss's backward formula is

$$y_s = y_3 + \Delta y_2 \binom{s}{1} + \Delta^2 y_2 \binom{s+1}{2} + \Delta^3 y_1 \binom{s+1}{3} + \Delta^4 y_1 \binom{s+2}{4} + \Delta^5 y_0 \binom{s+2}{5} + \Delta^6 y_0 \binom{s+3}{6} .$$

In both of these $s = (x - x_3)/h$. The forward formula uses the differences in the lower zigzag path, the backward formula, those in the upper.

Since each formula represents the interpolating polynomial for the values $y_0, y_1, \ldots, y_6$, all four must represent exactly the same polynomial, in spite of the difference in appearance. The expression for the error term is of course the same as that obtained in Chapter III, and is therefore

$$E = \frac{f^{(7)}(z)}{7!} (x - x_0)(x - x_1) \ldots (x - x_6)$$

The statement is frequently made in the literature on finite differences that central formulas such as Gauss's are more accurate than Newton's formulas. But if both have exactly the same expression for the error and in fact are really identical polynomials how can one be more accurate than another? What is really meant may be made clear by a glance at the graph of the function $(x - x_0)(x - x_1) \ldots (x - x_6)$, which is a factor of E. It is evident that if $f^{(7)}(z)$ does not change materially in the interval under consideration, the maximum error will be less in the interval $x_2 < x < x_4$ than in either of the intervals $x_0 < x < x_1$ or $x_5 < x < x_6$. In actual practice, Newton's forward formula would be used for the interval $x_0 < x < x_1$, Newton's backward for the interval $x_5 < x < x_6$ and one of Gauss's formulas in the interval $x_2 < x < x_4$. Hence, Gauss's formulas are not inherently more accurate than Newton's but are generally applied in an interval where the error is less.

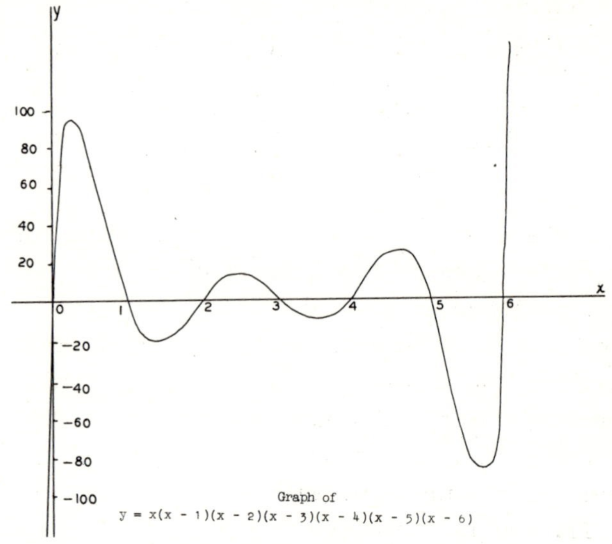

Graph of
$y = x(x - 1)(x - 2)(x - 3)(x - 4)(x - 5)(x - 6)$

Example 1. From the values of sin x at 10-degree intervals calculate sin 7°.

The table of values and the difference array appears as follows:

x	$y = \sin x$	Δy	$\Delta^2 y$	$\Delta^3 y$	$\Delta^4 y$	$\Delta^5 y$
0°	.00000					
		17365				
10°	.17365		- 528			
		16837		-511		
20°	.34202		-1039		31	
		15798		-480		14
30°	.50000		-1519		45	
		14279		-435		18
40°	.64279		-1954		63	
		12325		-372		2
50°	.76604		-2326		65	
		9999		-307		21
60°	.86603		-2633		86	
		7366		-221		-4
70°	.93969		-2854		82	
		4512		-139		
80°	.98481		-2993			
		1519				
90°	1.00000					

Since 7° lies near the beginning of the table, Newton's forward formula is appropriate. We have $h = 10$, $x_0 = 0$, $x = 10s$, so that for $x = 7^\circ$ we have $s = 0.7$. Newton's forward formula, using the top diagonal, is

$$y_s = 0 + .17365s - .00528\binom{s}{2} - .00511\binom{s}{3} + .00031\binom{s}{4} + .00014\binom{s}{5}.$$

Referring to Table II for $s = .7$ we obtain the values of $\binom{s}{2}$, $\binom{s}{3}$, $\binom{s}{4}$, and $\binom{s}{5}$ and substitute them in the formula above. Then

$$y = (.17365)(.7) + (.00528)(.10500) - (.00511)(.04550)$$
$$- (.00031)(.02616) + (.00014)(.01727) = 0.12187.$$

Example 2. Using the difference table of Example 1, find $\sin 86^\circ$.

Since 86° occurs near the end of the table, we use Newton's backward formula. We have $h = 10$, $x_0 = 90$, $x - 90 = 10s$, so that for $x = 86^\circ$, $s = -0.4$. Then Newton's backward formula, using the bottom diagonal of the array, is

$$y = 1.00000 - (.01519)(.4) + (.02993)(.12000)$$
$$+ (.00139)(.06400) - (.00082)(.04160)$$
$$+ (.00004)(.02995) = 0.99757$$

The correct value to 5 places is 0.99756.

Example 3. From the difference table of Example 1, obtain $\sin 45^\circ$.

Here we may use any one of the four forms. Let us choose Gauss's forward formula. We take $x_0 = 40^\circ$, $h = 10$, $x - 40 = 10s$ so that for $x = 45^\circ$ we have $s = 0.5$. Then the formula is

$$y = .64279 + .12325s - .01954\frac{s(s - 1)}{2!} - .00372\frac{s(s^2 - 1)}{3!}$$
$$+ .00063\frac{s(s^2 - 1)(s - 2)}{4!} + .00002\frac{s(s^2 - 1)(s^2 - 4)}{5!}$$

As we cannot readily use Table II for these coefficients, we substitute $s = 0.5$ and calculate the coefficients, obtaining

$$y = .64279 + (.12325)(.50000) + (.01954)(.12500)$$
$$+ (.00063)(.02344) + (.00002)(.01172) = 0.70711.$$

Exercises

1. From a table of common logarithms write the values of log 2.0, log 2.1, log 2.2, log 2.3, log 2.4, log 2.5. Then find the log 2.03, log 2.07, log 2,15, by interpolation. Compare results with the table.
2. From a table of cosines write the values of the cosines of 20°, 30°, 40°, 50°, 60°, 70°. Then find the cosines of 24°, 32°, 27°. Compare results with the table.
3. From a table of sines in radian measure write the values of the sines of 0.1, 0.2, 0.3, 0.4, 0.5, 0.6, 0.7 radians. Then find by interpolation the sines of 0.12, 0.13, 0.15, 0.16 radians. Compare results with the table.
4. From the cubes of 3, 4, 5, 6, find by interpolation the cube of 3.25.
5. From a table of log sines write the value of log sin 15°, 20°, 25°, 30°, 35°, and then find log sin $16^\circ 30'$, $18^\circ 20'$, $19^\circ 30'$ by interpolation.
6. From a table take the common logs of 100, 110, 120, 130, 140, 150, 160, and find by interpolation the logs of 115, 112.5, 117.5. Check.
7. From the following table find f(0.113575):

x	f(x)
0.11	0.0123456
0.12	0.0234467
0.13	0.0345811
0.14	0.0457498
0.15	0.0569538

8. Find the equation of the polynomial curve through the points

x:	0	1	2	3	4	5	6	7	8	9
y:	105	0	-15	0	9	0	-15	0	105	384

9. Show that if the order of the values of y be reversed, the signs of every odd difference column will be reversed. In problem 6 above find by interpolation log 155, 142.

10. Using Sheppard's rules set up all possible formulas of type (1) carried to 5th differences and starting with y_0.
How many different formulas can be obtained in this way?
How many essentially distinct interpolating polynomials are represented?

50. CENTRAL DIFFERENCES. STIRLING'S FORMULA

The Δ -notation so far used for differences is awkward when applied to the central difference formulas about to be considered. To get around this difficulty Sheppard introduced two new symbols, δ and μ, defined by the following equations

$$\delta f(x) = f(x + \tfrac{1}{2}h) - f(x - \tfrac{1}{2}h),$$

$$\mu f(x) = \tfrac{1}{2}\left[f(x + \tfrac{1}{2}h) + f(x - \tfrac{1}{2}h)\right],$$

in which h is the length of the interval between equally spaced given values of the variable x. If as usual we denote the given x's by ... x_{-2}, x_{-1}, x_0, x_1, x_2, ... and the corresponding values of $y = f(x)$ by ... y_{-2}, y_{-1}, y_0, y_1, y_2, ..., it is natural to denote

$$f(x_{-2} + \tfrac{1}{2}h) \quad \text{by } y_{-\frac{3}{2}},$$

$$f(x_{-1} + \tfrac{1}{2}h) \quad \text{by } y_{-\frac{1}{2}},$$

$$f(x_0 + \tfrac{1}{2}h) \quad \text{by } y_{\frac{1}{2}},$$

etc.

Note that in general these are not given values of the function f(x). With this notation we may write

$$\delta y_0 = y_{\frac{1}{2}} - y_{-\frac{1}{2}},$$

$$\mu y_0 = \frac{1}{2}(y_{\frac{1}{2}} + y_{-\frac{1}{2}}),$$

$$\delta y_1 = y_{\frac{3}{2}} - y_{\frac{1}{2}},$$

$$\mu y_1 = \frac{1}{2}(y_{\frac{3}{2}} + y_{\frac{1}{2}}),$$

etc.

Obviously the numerical values of these expressions are not directly obtainable from the given set of values of $f(x)$. On the other hand we have

$$\delta y_{\frac{1}{2}} = y_1 - y_0 = \Delta y_0$$

$$\delta y_{\frac{3}{2}} = y_2 - y_1 = \Delta y_1$$

etc.

and these can be found directly. If we apply the operator δ followed by the operator μ we have

$$\begin{aligned}\mu\delta y_0 &= \mu(y_{\frac{1}{2}} - y_{-\frac{1}{2}}) \\ &= \frac{1}{2}(y_1 + y_0 - y_0 - y_{-1}) \\ &= \frac{1}{2}(\Delta y_0 + \Delta y_{-1}).\end{aligned}$$

In like manner it may be shown that

$$\begin{aligned}\delta^2 y_0 &= y_1 - 2y_0 + y_{-1} = \Delta^2 y_{-1} \\ \mu\delta^3 y_0 &= \frac{1}{2}(\Delta^3 y_{-1} + \Delta^3 y_{-2}) \\ \delta^4 y_0 &= \Delta^4 y_{-2} \\ \mu\delta^5 y_0 &= \frac{1}{2}(\Delta^5 y_{-2} + \Delta^5 y_{-3})\end{aligned}$$

$$\delta^6 y_0 = \Delta^6 y_{-3},$$

$$\delta^3 y_{-\frac{1}{2}} = \Delta^3 y_{-2}$$

$$\delta^3 y_{\frac{1}{2}} = \Delta^3 y_{-1}$$

$$\delta^3 y_{\frac{3}{2}} = \Delta^3 y_0$$

etc.

In fact if m and n integers and both are odd or both are even, then

$$\delta^m y_{\frac{n}{2}} = \Delta^m y_{\frac{n-m}{2}}$$

The familiar table of differences appears as follows when expressed in the present notation:

$$\begin{array}{cccccc}
x_{-2} & y_{-2} & & & & \\
 & & \delta y_{-\frac{3}{2}} & & & \\
x_{-1} & y_{-1} & & \delta^2 y_{-1} & & \\
 & & \delta y_{-\frac{1}{2}} & & \delta^3 y_{-\frac{1}{2}} & \\
x_0 & y_0 & & \delta^2 y_0 & & \delta^4 y_0 \\
 & & \delta y_{\frac{1}{2}} & & \delta^3 y_{\frac{1}{2}} & \\
x_1 & y_1 & & \delta^2 y_1 & & \\
 & & \delta y_{\frac{3}{2}} & & & \\
x_2 & y_2 & & & &
\end{array}$$

The symmetry of this set-up as contrasted with the lack of symmetry of the same table in the Δ-notation is an advantage of the central difference symbols.

Gauss's forward formula expressed with the new symbols is

$$y_s = y_0 + \delta y_{\frac{1}{2}} \binom{s}{1} + \delta^2 y_0 \binom{s}{2} + \delta^3 y_{\frac{1}{2}} \binom{s+1}{3} + \delta^4 y_0 \binom{s+1}{4} + \dots$$

and his backward formula is

$$y_s = y_0 + \delta y_{-\frac{1}{2}} \binom{s}{1} + \delta^2 y_0 \binom{s+1}{2} + \delta^3 y_{-\frac{1}{2}} \binom{s+1}{3}$$

$$+ \delta^4 y_0 \binom{s+2}{4} + \dots$$

Evidently both of these are lacking in symmetry. But if we add the two together and divide by 2 we get

$$(1) \quad y_s = y_0 + \tfrac{1}{2}(\delta y_{\frac{1}{2}} + \delta y_{-\frac{1}{2}}) \binom{s}{1}$$

$$+ \delta^2 y_0 \, \tfrac{1}{2} \left[\binom{s}{2} + \binom{s+1}{2} \right] + \tfrac{1}{2}(\delta^3 y_{\frac{1}{2}} + \delta^3 y_{-\frac{1}{2}}) \binom{s+1}{3}$$

$$+ \delta^4 y_0 \, \tfrac{1}{2} \left[\binom{s+1}{4} + \binom{s+2}{4} \right] + \dots .$$

Now

$$(2) \qquad \tfrac{1}{2}(\delta y_{\frac{1}{2}} + \delta y_{-\frac{1}{2}}) = \mu\delta y_0$$

$$\tfrac{1}{2}(\delta^3 y_{\frac{1}{2}} + \delta^3 y_{-\frac{1}{2}}) = \mu\delta^3 y_0$$

etc.,

while

$$\tfrac{1}{2} \left[\binom{s}{2} + \binom{s+1}{2} \right] = \frac{s^2}{2!}$$

$$\tfrac{1}{2} \left[\binom{s+1}{4} + \binom{s+2}{4} \right] = \frac{s^2(s^2-1)}{4!}$$

$$\tfrac{1}{2} \left[\binom{s+2}{6} + \binom{s+3}{6} \right] = \frac{s^2(s^2-1)(s^2-2^2)}{6!}$$

etc.,

and

$$\binom{s+1}{3} = \frac{s(s^2-1)}{3!}, \quad \binom{s+2}{5} = \frac{s(s^2-1)(s^2-2^2)}{5!} \text{ etc.}$$

With the aid of these relations equation (1) becomes the symmetric formula

$$(3) \quad y_s = y_0 + \mu\delta y_0 s + \delta^2 y_0 \frac{s^2}{2!} + \mu\delta^3 y_0 \frac{s(s^2-1)}{3!} + \delta^4 y_0 \frac{s^2(s^2-1)}{4!} + \mu\delta^5 y_0 \frac{s(s^2-1)(s^2-2^2)}{5!} + \delta^6 y_0 \frac{s^2(s^2-1)(s^2-2^2)}{6!} + \ldots .$$

<u>This is Stirling's Central Difference Interpolation Formula</u>. The differences appearing in (3) occur in the difference table as shown in the diagram below:

———		———		———
	———		———	
———		———		———
	$\delta y_{-\frac{1}{2}}$		$\delta^3 y_{-\frac{1}{2}}$	
y_0		$\delta^2 y_0$		$\delta^4 y_0$
	$\delta y_{\frac{1}{2}}$		$\delta^2 y_{\frac{1}{2}}$	
———		———		———
	———		———	
———		———		———

In actual calculation the mean values $\mu\delta y_0$, $\mu\delta^3 y_0$, ... are obtained from the difference table by use of equations (2).

For the reasons mentioned at the end of Article 49 a central difference formula in general gives more accurate results than Newton's formula and hence is to be preferred whenever it is applicable, that is, except near the beginning or end of a table of values. For this reason and also because of its symmetry and simplicity Stirling's formula is much used for interpolation with differences.

Example. From the data given in Example 1, Article 49, find sin 44° by Stirling's formula.

We have $s = 0.4$ $\qquad y_0 = 0.64279,$

$$\mu\delta y_0 = \tfrac{1}{2}(.14279 + .12325) = 0.13302$$

$$\delta^2 y_0 = -0.01954$$

$$\mu\delta^3 y_0 = -0.00403$$

$$\delta^4 y_0 = 0.00063$$

$$\mu\delta^5 y_0 = 0.00010$$

Substituting these values together with $s = 0.4$ into Stirling's formula we have

$$\begin{aligned} y_s &= 0.64279 + (0.13302)(0.4) + (-.01954)(0.08) \\ &\quad + (-0.00403)(-0.056) + (0.00063)(-0.0056) \\ &= 0.69466. \end{aligned}$$

Exercises

Verify the following symbolic relations.

1. $\Delta = \mu\delta + \frac{1}{2}\delta^2$
2. $\mu^2 = 1 + \frac{1}{4}\delta^2$
3. $\Delta^2 = \delta^2 + \mu\delta^3 + \frac{1}{2}\delta^4$
4. $\delta^2 = \dfrac{\Delta^2}{1 + \Delta}$
5. $\mu\delta = \dfrac{\Delta + \frac{1}{2}\Delta^2}{1 + \Delta}$
6. $\mu\delta^3 = \dfrac{\Delta^3 + \frac{1}{2}\Delta^4}{(1 + \Delta)^2}.$

Using the data of Example 1, Article 49, and

Stirling's formula calculate

7. $\sin 25^{\circ}$

8. $\sin 37^{\circ}$

9. $\sin 52^{\circ}\ 30'$

10. If $\sin x = 0.90000$ find x. Use Stirling's formula together with the method of successive approximations given in Article 22.

51. EVERETT'S CENTRAL DIFFERENCE FORMULA

Gauss's forward formula is

$$y_s = y_0 + \delta y_{\frac{1}{2}} s + \delta^2 y_0 \binom{s}{2} + \delta^3 y_{\frac{1}{2}} \binom{s+1}{3}$$

$$+ \delta^4 y_0 \binom{s+1}{4} + \delta^5 y_{\frac{1}{2}} \binom{s+2}{5} + \delta^6 y_0 \binom{s+2}{6} + \cdots$$

In this we make the substitutions

$$\delta y_{\frac{1}{2}} = y_1 - y_0$$

$$\delta^3 y_{\frac{1}{2}} = \delta^2 y_1 - \delta^2 y_0$$

$$\delta^5 y_{\frac{1}{2}} = \delta^4 y_1 - \delta^4 y_0$$

$$\cdots\cdots\cdots$$

and

$$\binom{s}{2} = \binom{s+1}{3} - \binom{s}{3}$$

$$\binom{s+1}{4} = \binom{s+2}{5} - \binom{s+1}{5}$$

$$\binom{s+2}{6} = \binom{s+3}{7} - \binom{s+2}{7}$$

$$\cdots\cdots\cdots$$

The result can be arranged in the form

$$y_s =$$

$$(1 - s)y_0 - \delta^2 y_0\binom{s}{3} - \delta^4 y_0\binom{s+1}{5} - \delta^6 y_0\binom{s+2}{7} - \cdots$$

$$+ sy_1 + \delta^2 y_1\binom{s+1}{3} + \delta^4 y_1\binom{s+2}{5} + \delta^6 y_1\binom{s+3}{7} + \cdots$$

This equation may be put in a still more attractive form by letting $1 - s = t$ in the first line and noting that

$$\binom{s}{3} = \binom{1-t}{3} = \frac{(1-t)(-t)(-1-t)}{6} = -\binom{t+1}{3}$$

and similarly

$$\binom{s+1}{5} = -\binom{t+2}{5}, \quad \binom{s+2}{7} = -\binom{t+3}{7}$$

so that we have finally <u>Everett's formula</u>

(1) $$y_s =$$

$$ty_0 + \binom{t+1}{3}\delta^2 y_0 + \binom{t+2}{5}\delta^4 y_0 + \binom{t+3}{7}\delta^6 y_0 + \cdots$$

$$+ sy_1 + \binom{s+1}{3}\delta^2 y_1 + \binom{s+2}{5}\delta^4 y_1 + \binom{s+3}{7}\delta^6 y_1 + \cdots$$

where $t = 1 - s$. This is a remarkably elegant and useful interpolation formula. It utilizes the even ordered central differences in two adjacent rows of the difference table as exhibited in the scheme below:

—		—		—		—
	—		—		—	
y_0		$\delta^2 y_0$		$\delta^4 y_0$		$\delta^6 y_0$
	—		—		—	
y_1		$\delta^2 y_1$		$\delta^4 y_1$		$\delta^6 y_1$
	—		—		—	
—		—		—		—

The coefficients up to the fifth degree are given in Table III.

The principal use of Everett's formula is for interpolation in published tables in which the even-ordered differences are given in addition to the tabular values. In such a table if sixth differences are negligible, for example, only the second and fourth differences need to be printed, requiring two columns instead of the five columns needed for all the first five differences. The resulting reduction of printing costs makes this procedure most desirable.

Example. The table of the function Si(x) prepared by the Federal Works Agency contains the following entries:

x	Si(x)	δ^2Si(x)
59.52	1.5872974053	-16606
59.53	1.5873250422	-16628

Fourth differences are negligible. Find the value of Si(x) for $x = 59.5245$.

Here $s = 0.45$, $t = 0.55$, $\binom{s+1}{3} = -0.05981$, $\binom{t+1}{3} = -0.06394$ from Table III. Then

$$\begin{aligned} Si(x) &= 1.5872974053\ (0.55) - (0.0000016606)(-0.06394) \\ &\quad + 1.5873250422\ (0.45) - (0.0000016628)(-0.05981) \\ &= 1.5873100475. \end{aligned}$$

Everett's formula also is particularly well adapted to subtabulation, as will be shown in Article 54.

52. BESSEL'S FORMULA

Another central difference formula can be derived in the following manner. Gauss's backward formula starting at y_1 instead of y_0 can be written

$$y_s = y_1 + \delta y_{\frac{1}{2}} \binom{s-1}{1} + \delta^2 y_1 \binom{s}{2} + \delta^3 y_{\frac{1}{2}} \binom{s}{3} + \delta^4 y_1 \binom{s+1}{4} + \ldots$$

with $s = (x - x_0)/h$. Gauss's forward formula starting at y_0 with the same definition of s is

$$y_s = y_0 + \delta y_{\frac{1}{2}} \binom{s}{1} + \delta^2 y_0 \binom{s}{2} + \delta^3 y_{\frac{1}{2}} \binom{s+1}{3} + \delta^4 y_0 \binom{s+1}{4} + \ldots .$$

If we add these together and divide by 2 we have

$$y_s = \mu y_{\frac{1}{2}} + \delta y_{\frac{1}{2}} \left(\frac{2s-1}{2}\right) + \mu\delta^2 y_{\frac{1}{2}} \binom{s}{2}$$

$$+ \delta^3 y_{\frac{1}{2}} \frac{s(s-1)(2s-1)}{2 \cdot 3!} + \mu\delta^4 y_{\frac{1}{2}} \binom{s+1}{4} + \ldots .$$

The result may be put in a more attractive form by the substitution $s = t + \frac{1}{2}$, for we then have

$$y_{t+\frac{1}{2}} = \mu y_{\frac{1}{2}} + \delta y_{\frac{1}{2}} t + \mu\delta^2 y_{\frac{1}{2}} \frac{(t^2 - \frac{1}{4})}{2!} + \delta^3 y_{\frac{1}{2}} \frac{t(t^2 - \frac{1}{4})}{3!}$$

$$+ \mu\delta^4 y_{\frac{1}{2}} \frac{(t^2 - 1/4)(t^2 - 9/4)}{4!} + \ldots .$$

This is <u>Bessel's Formula</u>. The differences utilized appear in the difference array as shown below

———		———		———
y_0	———	$\delta^2 y_0$	———	$\delta^4 y_0$
	$\delta y_{\frac{1}{2}}$		$\delta^3 y_{\frac{1}{2}}$	
y_1		$\delta^2 y_1$		$\delta^4 y_1$
	———		———	
———		———		———

<u>Exercises</u>

1. From a table of tangents obtain the tangents of 0°, 10°, 20°, 30°, 40°, 50°, 60°. Construct a difference table for these values.
2. With the result of exercise 1 find tan 33° by Stirling's formula.

Exercises

3. Find tan 35° by Everett's formula.
4. Find tan 36° by Bessel's formula.
5. Find tan 28° by Stirling's formula.
6. In Bessel's formula set $t = 0$ and obtain a formula for interpolation at the midpoint of an interval.

53. TABULATION OF POLYNOMIALS

The method of differences provides a convenient means for building up a table of values for a given polynomial P(x) at equally spaced values of x. The procedure will be illustrated by an example.

Example 1. Construct a table of values for

$$y = 2x^4 - 3x^3 + 5x^2 - 2x + 3$$

starting with $x = 1$ and proceeding at intervals of 0.1.

Since the polynomial is of degree 4 the 4th differences are constant. Hence, we first calculate five values by direct substitution and form the differences

x	y	Δy	$\Delta^2 y$	$\Delta^3 y$	$\Delta^4 y$
1	5.0000				
		7852			
1.1	5.7852		1928		
		9780		372	
1.2	6.7632		2300		48
		12080		420	
1.3	7.9712		2720		
		14800			
1.4	9.4512				

Since the 4th difference will be constant the next line of differences can be formed from right to left as follows:

1.5	11.2500	17988	3188	468	48

then the next line is calculated, etc. An alternative method is to complete the column of 4th differences, then by repeated additions complete the column of 3rd difference etc. This has some advantages where the differences are not all positive. Whichever method is used, the result should be checked from time to time with values calculated directly from the original polynomial. The computation to $x = 2$ appears as follows

					48
				420	
			2720		48
		14800		468	
1.4	9.4512		3188		48
		17988		516	
1.5	11.2500		3704		48
		21692		564	
1.6	13.4192		4268		48
		25960		612	
1.7	16.0152		4880		48
		30840		660	
1.8	19.0992		5540		48
		36380		708	
1.9	22.7372		6248		
		42628			
2.0	27.0000				

The value for $x = 2.0$ checks with that given by direct substitution.

In the foregoing example the functional values were exact, no significant figures having been dropped. In practice, however, the values are ordinarily not exact, being rounded off after a specified number of decimal places, and in consequence the tabulated n-th differences of a polynomial of n-th degree will not be constant. In such cases the cumulative effect of small errors in the differences may seriously affect the accuracy of the y itself.

In a case of this kind the method of the following Article may be used.

54. SUBTABULATION

In the construction of tables of a function $y = f(x)$ it is the usual practice not to compute directly every entry in the final table, but to compute a primary table with fewer entries (say every tenth entry of the final table) and then to fill in the remainder by interpolation.

One way of filling in such a table is to calculate each of the subtabulated values directly by interpolation. For this purpose Everett's formula offers peculiar advantages. Let

$$F_i(s) = sy_i + \binom{s+1}{3} \delta^2 y_i + \binom{s+2}{5} \delta^4 y_i + \ldots$$

Then in the interval between:

$$\begin{array}{ll} x_0 \text{ and } x_1 & y_s = F_0(t) + F_1(s) ; \\ x_1 \text{ and } x_2 & y_s = F_1(t) + F_2(s) ; \\ x_2 \text{ and } x_3 & y_s = F_2(t) + F_3(s) ; \end{array}$$

and so on. Since $t = 1 - s$ the set of values obtained for $F_1(t)$ will be the same as those for $F_1(s)$ but in the reverse order. The same holds for $F_2(t)$ and $F_2(s)$. In an extended calculation this duplication reduces the total labor of interpolation by nearly one half.

<u>Example 1</u>. In the following table of sin x

x	y	$\delta^2 y$	$\delta^4 y$
0.2	.19867	-199	4
0.3	.29552	-295	2
0.4	.38942	-389	3

supply subtabulated values at intervals of 0.02.

Using Table III for the values of the coefficients we calculate $F_i(.2)$, $F_i(.4)$, $F_i(.6)$, $F_i(.8)$ for $i = 0, 1, 2$, and arrange the computation as shown:

x	y				
0.20	.19867	t	$F_0(t)$	s	$F_1(s)$
0.22	.21823	.8	.159032	.2	.059199
0.24	.23770	.6	.119330	.4	.118373
0.26	.25708	.4	.079580	.6	.177501
0.28	.27636	.2	.039798	.8	.236558
0.30	.29552	t	$F_1(t)$	s	$F_2(s)$
0.32	.31457	.8	.236558	.2	.078009
0.34.	.33349	.6	.177501	.4	.155986
0.36	.35227	.4	.118373	.6	.233901
0.38	.37092	.2	.059199	.8	.311723
0.40	.38942				

The columns for $F_0(t)$, $F_1(s)$, etc., are computed directly and the final entries for y are then obtained by addition. The extra decimal place is retained as a guard.

Instead of calculating directly each subtabulated value as explained above a very common procedure is to determine a row of differences for the subtabulated values and then to build up the values themselves exactly as in Article 53, on the assumption that differences above a certain order vanish. We now explain how this may be done.

54. SUBTABULATION

Let the interval for the primary table be H and denote the difference operator for this interval by Δ_H. Let h and Δ_h be similarly defined for the final table. Using Newton's interpolation formula in the symbolic form we have

$$y(x) = (1 + \Delta_H)^S y_0 = (1 + \Delta_h)^s y_0$$

in which $S = \dfrac{x - x_0}{H}$, $s = \dfrac{x - x_0}{h}$. Since the two operators produce the same result they are symbolically equivalent so that

$$(1 + \Delta_H)^S = (1 + \Delta_h)^s.$$

from which by inserting the given values of S and s and solving for Δ_h we obtain

$$\Delta_h = (1 + \Delta_H)^{h/H} - 1 \tag{1}$$

or in expanded form

$$\Delta_h = \binom{h/H}{1} \Delta_H + \binom{h/H}{2} \Delta^2{}_H + \binom{h/H}{3} \Delta^3{}_H + \ldots$$

Also

$$\Delta^2{}_h = \left[(1 + \Delta_H)^{h/H} - 1 \right]^2, \tag{2}$$

$$\Delta^3{}_h = \left[(1 + \Delta_H)^{h/H} - 1 \right]^3, \tag{3}$$

etc.

These relationships make it possible to compute differences in the final table from the differences constructed for the primary table. We then build up the missing entries by the method of the preceding article.

Example 2. From a table calculated for $y = (x^2 - 1)(x^2 - 4)/4$ at intervals of 0.1 obtain a table with intervals of 0.01.

The first six entries of the primary table together with differences are given below

x	y	$\Delta_H y$	$\Delta^2_H y$	$\Delta^3_H y$	$\Delta^4_H y$
0	1.000000				
		-12475			
.1	.987525		-24650		
		-37125		900	
.2	.950400		-23750		600
		-60875		1500	
.3	.889525		-22250		600
		-83125		2100	
.4	.806400		-20150		
		-103275			
.5	.703125				

For this case h/H = 0.1, and equations (1), (2), (3), etc., in expanded form turn out to be

$$\Delta_h = 0.1\ \Delta_H - 0.045\ \Delta^2_H + 0.0285\ \Delta^3_H - 0.02066\ \Delta^4_H + \cdots$$

$$\Delta^2_h = 0.01\ \Delta^2_H - 0.009\ \Delta^3_H + 0.007725\ \Delta^4_H - \cdots$$

$$\Delta^3_h = 0.001\ \Delta^3_H - 0.00135\ \Delta^4_H + \cdots$$

$$\Delta^4_h = 0.0001\ \Delta^4_H - \cdots$$

Using these formulas together with the primary table we compute the forward differences for the entries x = 0, and x = 0.1 and record them in their proper places in the final table. At this stage the latter appears as follows:

x	y	$\Delta_h y$	$\Delta^2_h y$	$\Delta^3_h y$	$\Delta^4_h y$
0	1.000000				
		-125.0			
0.01			-250.0		
				0.09	
2					0.06
3					
4					
5					
6					
7					
8					
9					
10	.987525				
		-2613.4			
11			-246.4		
				0.69	
12					0.06
13					
14					

The decimals in the differences indicate figures beyond the sixth place which are retained to guard against accumulated errors. Even with this precaution it is usually necessary to carry the primary table to a greater number of places than are to be retained in the final table.

It is now a simple matter to fill in the block of the final table from $x = 0$ to $x = 0.1$. The result appears below:

x	y	$\Delta_h y$	$\Delta^2_h y$	$\Delta^3_h y$	$\Delta^4_h y$
0	1.000000				
		-125.0			
0.01	.999875		-250.0		
		-375.0		0.09	
2	.999500		-249.9		0.06
		-624.9		.15	
3	.998875		-249.8		0.06
		-874.7		.21	
4	.998000		-249.5		0.06
		-1124.2		.27	
5	.996876		-249.3		0.06
		-1373.5		.33	
6	.995503		-249.0		0.06
		-1622.1		.39	
7	.993881		-248.6		0.06
		-1870.7		.45	
8	.992010		-248.1		0.06
		-2118.8		.51	
9	.989891		-247.6		0.06
		-2366.4		.57	
10	.987525		-247.0		0.06
		-2613.4		.63	
			-246.4		0.06
				.69	
					0.06

To reduce the danger of accumulated errors, the entries for $x = 0.06, 7, 8, 9$ were calculated by working backward from $x = 0.1$.

This example gives a general idea of the method of

subtabulation. Actually many variations in procedure are possible. Central differences may be used instead of forward differences. Or a few values in each block of the final table can be calculated by interpolation, the difference table formed, and the remaining values in the block filled in by continuing the differences. This is especially convenient if the third and higher order differences in the final table are negligible.

<u>Exercises</u>

From the table below construct a table to six places at intervals of 0.001.

x	y
.0	1.5707963
.01	1.5747456
.02	1.5787399
.03	1,5827803
.04	1.5868678
.05	1.5910035

55. DERIVATIVES IN TERMS OF DIFFERENCES

From the interpolation formulas in terms of differences it is easy to derive formulas for numerical differentiation expressed in terms of differences. For example by differentiating Stirling's formula

$$y(x) = y_0 + \mu\delta y_0 s + \delta^2 y_0 \frac{s^2}{2!} + \mu\delta^3 y_0 \frac{s(s^2-1)}{3!} + \delta^4 y_0 \frac{s^2(s^2-1)}{4!} + \ldots,$$

recalling that $dx = h\,ds$, and afterwards setting $s = 0$, we have

(1) $$\frac{dy}{dx}\bigg|_{x=x_0} = \frac{\mu}{h}\left[\delta y_0 - \frac{1^2}{3!}\delta^3 y_0 + \frac{1^2\cdot 2^2}{5!}\delta^5 y_0 - \frac{1^2\cdot 2^2\cdot 3^2}{7!}\delta^7 y_0 + \cdots\right]$$

Similarly

(2) $$\frac{d^2y}{dx^2}\bigg|_{x=x_0} = \frac{1}{h^2}\left[\delta^2 y_0 - \frac{2}{4!}\delta^4 y_0 + \frac{2\cdot 2^2}{6!}\delta^6 y_0 - \frac{2\cdot 2^2\cdot 3^2}{8!}\delta^8 y_0 + \cdots\right]$$

Newton's Forward Difference Interpolation Formula in symbolic form is

$$y(x) = (1 + \Delta)^s y_0$$

with the relation $x - x_0 = hs$. Differentiation of this equation with respect to x gives formally

(4) $$dy/dx = \frac{1}{h}(1 + \Delta)^s \ln(1 + \Delta)y_0,$$

or, after expansion in powers of Δ,

(5) $$dy/dx = \frac{1}{h}\left\{\Delta y_0 + \left[s - \frac{1}{2}\right]\Delta^2 y_0 + \left[\binom{s}{2} - \frac{1}{2}s + \frac{1}{3}\right]\Delta^3 y_0 + \left[\binom{s}{3} - \frac{1}{2}\binom{s}{2} + \frac{1}{3}s - \frac{1}{4}\right]\Delta^4 y_0 + \cdots\right\}$$

It can be shown that if the function y(x) is a polynomial and consequently differences from some point on all vanish so that questions of convergence do not arise then formula

(4) is actually valid. Also if desired (5) can be obtained by direct differentiation of Newton's forward difference interpolation formula without any use of the symbolic operation $(1 + \Delta)^s$.

In particular if $x = x_0$, $s = 0$, equations (4) and (5) reduce to

$$(6) \qquad (dy/dx)_{x=x_0} = \frac{1}{h}(\Delta y_0 - \frac{1}{2}\Delta^2 y_0 + \frac{1}{3}\Delta^3 y_0 - \frac{1}{4}\Delta^4 y_0 + \ldots)$$

Example. From the table of difference for $y = \sin x$ given in Article 45, calculate dy/dx at $x = 0.2$.

The forward differences at $x = 0.2$ given in the table are $\Delta y = 0.09685$, $\Delta^2 y = -0.00295$, $\Delta^3 y = -0.00094$, $\Delta^5 y = 0.00003$. With these values formula (6) gives

$$dy/dx = 10(0.09685 + 0.001475 - 0.000313 - 0.000007)$$
$$= 0.98005.$$

The correct result should be 0.98007.

Exercises

1. From the table in Article 45 obtain dy/dx at $x = 0.2$ using formula (1).
2. Obtain dy/dx at $x = 0.04$ using (5).
3. Obtain d^2y/dx^2 at $x = 0.3$ using (2).

56. INTEGRAL OF NEWTON'S INTERPOLATION FORMULA

Integrating the formula

$$y(x) = (1 + \Delta)^s y_k$$

in which $x - x_k = hs$ and $dx = hds$, we have formally

$$(1) \quad \int_{x_k}^{x} y(x)\,dx = h \int_0^s (1 + \Delta)^s\,ds\; y_k = h\frac{(1 + \Delta)^s - 1}{\ln(1 + \Delta)} y_k$$

Now the quantity $\Delta/\ln(1 + \Delta)$ may be expanded in a series

(2) $$\frac{\Delta}{\ln(1+\Delta)} = 1 + \frac{1}{2}\Delta - \frac{1}{12}\Delta^2 + \frac{1}{24}\Delta^3 - \frac{19}{720}\Delta^4 + \frac{3}{160}\Delta^5 - \frac{863}{60480}\Delta^6 + \frac{275}{24192}\Delta^7 - \frac{33953}{3628800}\Delta^8 + \dots .$$

We also have the expansion

(3) $$\frac{(1+\Delta)^s - 1}{\Delta} = s + \Delta\binom{s}{2} + \Delta^2\binom{s}{3} + \Delta^3\binom{s}{4} + \dots$$

If for brevity the coefficient of Δ^i in (2) be designated by g_i equation (1) can be put in the form

(4) $$\int_{x_k}^{x} y(x)\,dx = h\Big\{ sy_k + \left[g_1 s + g_0\binom{s}{2}\right]\Delta y_k + \left[g_2 s + g_1\binom{s}{2} + g_0\binom{s}{3}\right]\Delta^2 y_k + \dots + \sum_{i=0}^{m} g_{m-i}\binom{s}{i+1}\Delta^m y_k + \dots\Big\} .$$

This equation is valid if $y(x)$ is any polynomial. From it may be obtained a number of useful formulas. For example, if $x = x_{k+1}$, $s = 1$, $\binom{s}{i} = 0$ if $i > 1$ and we get

(5) $$\int_{x_k}^{x_{k+1}} y(x)\,dx = h\sum_i g_i\,\Delta^i y_k .$$

Again if $x = x_{k+2}$, $s = 2$, and

$$(6)\qquad \int_{x_k}^{x_{k+2}} y(x)\,dx = h \sum_i (2g_i + g_{i-1})\,\Delta^i y_k .$$

The corresponding formulas for backward differences are

$$(5')\qquad \int_{x_{k-1}}^{x_k} y(x)\,dx = h \sum_i g_i (-1)^i\,\Delta^i y_{k-i}$$

and

$$(6')\qquad \int_{x_{k-2}}^{x_k} y(x)\,dx = h \sum_i (2g_i + g_{i-1})(-1)^i\,\Delta^i y_{k-i}$$

If in (5) we let $k = 0, 1, \ldots, n - 1$, in succession and sum the resulting equations we derive the formula

$$(7)\qquad \int_{x_0}^{x_n} y(x)\,dx = h(\tfrac{1}{2}y_0 + y_1 + y_2 + y_3 + \cdots + y_{n-1} + \tfrac{1}{2}y_n)$$

$$- \frac{h}{12}(\Delta y_n - \Delta y_0) + \frac{h}{24}(\Delta^2 y_n - \Delta^2 y_0) - \cdots$$

$$- \frac{19h}{720}(\Delta^3 y_n - \Delta^3 y_0) + \frac{3h}{160}(\Delta^4 y_n - \Delta^4 y_0) + \cdots .$$

This formula is in effect the trapezoidal rule with correction terms expressed by means of differences. The differences involved lie in parallel diagonals in the difference table sloping downward to the right. Obviously, (7) involves ordinates lying outside the range of integration. By successive substitutions into (7) from formulas of the type

$$\Delta^k y_m = \Delta^k y_{m-1} + \Delta^{k+1} y_{m-1}$$

equation (7) can be transformed into

$$(8)\quad \int_{x_0}^{x_n} y(x)\,dx = h(\tfrac{1}{2}y_0 + y_1 + y_2 + \cdots + y_{n-1} + \tfrac{1}{2}y_n)$$

$$- \frac{h}{12}(\Delta y_{n-1} - \Delta y_0) - \frac{h}{24}(\Delta^2 y_{n-2} + \Delta^2 y_0)$$

$$- \frac{19h}{720}(\Delta^3 y_{n-3} - \Delta^3 y_0) - \frac{3h}{160}(\Delta^4 y_{n-4} + \Delta^4 y_0) - \cdots$$

which employs differences in the diagonal forward from y_0 and in the diagonal backward from y_n. If differences beyond the nth order are ignored, formula (8) involves only ordinates in the range of integration. This is Gregory's Formula.

57. SYMMETRIC INTEGRALS OF STIRLING'S FORMULA

In Stirling's formula, it will be noted that the coefficients of the odd-ordered differences are odd functions of s. Hence, if Stirling's formula is integrated between symmetric limits $-s$ to $+s$ all of the differences of odd order drop out. In this way we obtain the set of symmetric formulas

$$(1)\quad \int_{x_{-1}}^{x_1} y(x)\,dx = h\Big[2y_0 + \frac{1}{3}\delta^2 y_0 - \frac{1}{90}\delta^4 y_0 + \frac{1}{756}\delta^6 y_0$$

$$- \frac{23}{113400}\delta^8 y_0 + \frac{263}{7484400}\delta^{10} y_0 - \cdots\Big]$$

$$(2)\quad \int_{x_{-2}}^{x_2} y(x)\,dx = h\Big[4y_0 + \frac{8}{3}\delta^2 y_0 + \frac{14}{45}\delta^4 y_0 - \frac{8}{945}\delta^6 y_0$$

$$+ \frac{13}{14175}\delta^8 y_0 - \frac{62}{467775}\delta^{10} y_0 + \cdots\Big]$$

$$(3)\quad \int_{x_{-3}}^{x_3} y(x)\,dx = h\left[6y_0 + \frac{27}{3}\delta^2 y_0 + \frac{33}{10}\delta^4 y_0 + \frac{41}{140}\delta^6 y_0 - \frac{9}{1400}\delta^8 y_0 + \frac{19}{30800}\delta^{10} y_0 - \cdots\right]$$

$$(4)\quad \int_{x_{-4}}^{x_4} y(x)\,dx = h\left[8y_0 + \frac{64}{3}\delta^2 y_0 + \frac{688}{45}\delta^4 y_0 + \frac{736}{189}\delta^6 y_0 + \frac{3956}{14175}\delta^8 y_0 - \frac{2368}{467775}\delta^{10} y_0 + \cdots\right]$$

$$(5)\quad \int_{x_{-5}}^{x_5} y(x)\,dx = h\left[10y_0 + \frac{125}{3}\delta^2 y_0 + \frac{875}{18}\delta^4 y_0 + \frac{17225}{756}\delta^6 y_0 + \frac{20225}{4536}\delta^8 y_0 + \frac{80335}{299376}\delta^{10} y_0 - \cdots\right]$$

Exercises

Show how formulas (2), (4), (6), and (8) of Article 34 may be derived from (1), (2), (3), (4), respectively, above. For example in (1) above replace $\delta^2 y_0$ by its value $y_1 - 2y_0 + y_{-1}$ and obtain

$$\int_{x_{-1}}^{x_1} y(x)\,dx = \frac{h}{3}\left[y_1 + 4y_0 + y_{-1}\right] - \frac{h}{90}\delta^4 y_0 + \cdots .$$

Except for notation this is (2) of Article 34.

58. INTEGRAL OF EVERETT'S FORMULA

Everett's formula is

$$y(x) = y_0 t + \delta^2 y_0 \binom{t+1}{3} + \delta^4 y_0 \binom{t+2}{5} + \delta^6 y_0 \binom{t+3}{7}$$

$$+ \cdot \cdot \cdot + y_1 s + \delta^2 y_1 \binom{s+1}{3} + \delta^4 y_1 \binom{s+2}{5} + \delta^6 y_1 \binom{s+3}{7}$$

$$+ \cdot \cdot \cdot$$

with $t = (x_1 - x)/h$, $s = (x - x_0)/h$. Since

$$\int_{x_0}^{x_1} dx = h \int_0^1 ds = h \int_1^0 (-dt) = h \int_0^1 dt$$

we have

$$(1) \qquad \int_{x_0}^{x_1} y(x)\, dx = \frac{h}{2}(y_0 + y_1) - \frac{h}{24}(\delta^2 y_0 + \delta^2 y_1)$$

$$+ \frac{11h}{1440}(\delta^4 y_0 + \delta^4 y_1)$$

$$- \frac{191h}{120960}(\delta^6 y_0 + \delta^6 y_1) + \frac{2497h}{7257600}(\delta^8 y_0 + \delta^8 y_1) - \cdot \cdot \cdot \cdot$$

If we add together formula (1) and the formulas obtained by advancing the subscripts $1, 2, \ldots, n-1$ we have

$$(2) \qquad \int_{x_0}^{x_n} y(x)\, dx = h(\tfrac{1}{2}y_0 + y_1 + y_2 + \cdot \cdot \cdot + y_{n-1} + \tfrac{1}{2}y_n)$$

$$- \frac{h}{24}(\Delta y_{n-1} + \Delta y_n - \Delta y_{-1} - \Delta y_0)$$

$$+ \frac{11h}{1440}(\Delta^3 y_{n-2} + \Delta^3 y_{n-1} - \Delta^3 y_{-2} - \Delta^3 y_{-1})$$

This is more rapidly convergent than Gregory's formula. On the other hand, it has the disadvantage of using ordinates outside the range of integrations. It is interesting to note that (2) contains formulas (4) and (5) of Article 32 as special cases.

Example 1. Using differences find $\int_0^{0.6} \sin x \, dx$.

Method 1. Using the table of differences p. 156 and Gregory's formula, Art. 56 formula (8), we have

$$\int_0^{0.6} \sin x \, dx = 0.1(1.74519)$$

$$- \frac{0.1}{12}(0.08521 - 0.09983)$$

$$- \frac{0.1}{24}(-0.00480 - 0.00099)$$

$$- \frac{1.9}{720}(-0.00091 + 0.00100)$$

$$= 0.174665$$

Method 2. Using formula (3), Art. 57, with the subscripts advanced by 3 we get

$$\int_0^{0.6} \sin x \, dx = 0.1 \; 6(0.29552) + \frac{27}{3}(-0.00295)$$

$$+ \frac{33}{10}(0.00002) = 0.174664.$$

Example 2. Find $\int_{0.3}^{0.4} \sin x \, dx$. Here let us try (1), Art. 58. This gives

$$\int_{0.3}^{0.4} \sin x\, dx = \frac{0.1}{2}(0.29552 + 0.38942) - \frac{0.1}{24}(-0.00295$$

$$- 0.00389) = 0.034275.$$

Exercises

1. Integrate $\frac{1}{1 + x^4}$ from 0 to 1, using 10 intervals, by Gregory's formula.

2. Integrate $\sqrt{1 + x^2}$ from 0 to 1.2, using 6 intervals, by (3) Art. 57.

3. Integrate $\frac{\sin x}{x}$ from 0 to 1, using 5 intervals, by (2) Art. 58.

4. Integrate $\frac{e^x}{x}$ from 1 to 3, using 4 intervals, by Gregory's formula.

5. Integrate $\frac{e^x}{x}$ from 1 to 3, using 4 intervals, by (2) Art. 57.

6. $\int_0^1 \sqrt[3]{1 + x^2}\, dx$

7. $\int_0^{\pi} \sin x^2\, dx$

8. $\int_0^4 e^{-x^2}\, dx$

9. $\int_0^1 \sqrt{1 + x^3}\, dx$

10. $\int_{-1}^1 \sqrt{1 - x^2}\, dx = \frac{\pi}{2}.$

Chapter VII
DIVIDED DIFFERENCES*

The method of differences developed in the preceding chapter is restricted to the case of equally spaced values of the dependent variable. The method of divided differences avoids this limitation, and hence is of much more general application.

59. DEFINITION OF DIVIDED DIFFERENCES

Let y(x) be any function of x having values $y_0, y_1, \ldots, y_n$ corresponding to $x_0, x_1, \ldots, x_n$ respectively. The fraction

$$\frac{y_i - y_j}{x_i - x_j}$$

will be denoted by the symbol† $[x_i x_j]$, so that

(1) $$[x_0 x_1] = \frac{y_0 - y_1}{x_0 - x_1}, \quad [x_1 x_2] = \frac{y_1 - y_2}{x_1 - x_2}, \text{ etc.}$$

Evidently it is also true that $[x_0 x_1] = \dfrac{y_1 - y_0}{x_1 - x_0}$ $= [x_1 x_0]$, so that the order of the quantities inside the brackets is immaterial. These fractions are called first order divided differences of the function y(x).

Using a similar notation we form the fractions

*See Appendix C.
†See Appendix A.

$$[x_0x_1x_2] = \frac{[x_0x_1] - [x_1x_2]}{x_0 - x_2},$$

(2)

$$[x_1x_2x_3] = \frac{[x_1x_2] - [x_2x_3]}{x_1 - x_3}$$

etc.

These are called <u>second order divided difference</u> of $y(x)$. Note that they are formed by dividing the difference of two first order differences having one x in common by the difference of the other two x's.

Similarly third order divided differences are obtained from second order divided differences by formulas of the type

$$(3) \qquad [x_0x_1x_2x_3] = \frac{[x_0x_1x_2] - [x_1x_2x_3]}{x_0 - x_3}$$

and likewise the divided differences of the nth order are obtained from those of $(n - 1)^{th}$ order by the formula

$$(4) \qquad [x_0x_1 \ldots x_n] = \frac{[x_0x_1 \ldots x_{n-1}] - [x_1x_2 \ldots x_n]}{x_0 - x_n}$$

Note that formulas (2), (3), and (4) remain valid if the order of subtraction in both numerator and denominator is reversed.

For the sake of harmony in the notation it is often convenient to write $[x_0] = y_0$, $[x_1] = y_1$, etc. In actual calculation it is customary to arrange the x's, y's, and the divided differences in columns according to the following scheme:

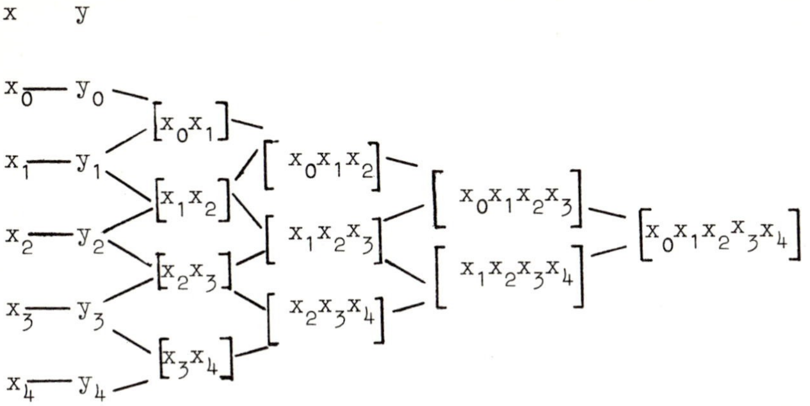

x	y				
x_0	y_0				
		$[x_0x_1]$			
x_1	y_1		$[x_0x_1x_2]$		
		$[x_1x_2]$		$[x_0x_1x_2x_3]$	
x_2	y_2		$[x_1x_2x_3]$		$[x_0x_1x_2x_3x_4]$
		$[x_2x_3]$		$[x_1x_2x_3x_4]$	
x_3	y_3		$[x_2x_3x_4]$		
		$[x_3x_4]$			
x_4	y_4				

<u>The value of any bracket in the table is found by dividing the difference between the two adjacent values to the left by the difference between the x's on the two diagonals through the desired bracket.</u> Thus

$$[x_1x_2x_3x_4] = \frac{[x_2x_3x_4] - [x_1x_2x_3]}{x_4 - x_1}$$

Note that in every instance the differences in the numerator have all x's but one in common, and the denominator is the difference of the x's not common to both.

A numerical example will show the simplicity of the calculation.

Example 1. Obtain the divided differences for the values

Xx	−3	−1	2	4	6	7
y	−1584	216	−144	96	−288	−504

We write the x's and y's in parallel columns and calculate the divided differences in succeeding columns according to the rule above. The completed table appears as follows:

x	y					
-3	-1584					
		900				
-1	216		-204			
		-120		36		
2	-144		48		-6	
		120		-18		1
4	96		-78		4	
		-192		14		
6	-288		-8			
		-216				
7	-504					

For instance to obtain the first entry in the first column of differences we have

$$\frac{(216) - (-1584)}{(-1)-(-3)} = 900.$$

To obtain the second entry in the third column of differences we have

$$\frac{(-78) - (48)}{6 - (-1)} = -18.$$

Divided differences may be expressed in a symmetric form that has important applications. It is evident that

$$\left[x_1 x_2\right] = \frac{y_1}{x_1 - x_2} + \frac{y_2}{x_2 - x_1},$$

$$\left[x_0 x_1\right] = \frac{y_0}{x_0 - x_1} + \frac{y_1}{x_1 - x_0}$$

and from these we get

$$[x_0x_1x_2] = \frac{[x_0x_1] - [x_1x_2]}{x_0 - x_2}$$

$$= \frac{1}{x_0 - x_2}\left[\frac{y_0}{x_0 - x_1} + \frac{[x_0 - x_2]\, y_1}{[x_1 - x_0]\quad[x_1 - x_2]} - \frac{y_2}{x_2 - x_1}\right]$$

$$= \frac{y_0}{(x_0-x_1)(x_0-x_2)} + \frac{y_1}{(x_1-x_0)(x_1-x_2)} + \frac{y_2}{(x_2-x_0)(x_2-x_1)}.$$

In the same way we find

$$[x_0x_1x_2x_3] = \frac{y_0}{(x_0-x_1)(x_0-x_2)(x_0-x_3)} + \frac{y_1}{(x_1-x_0)(x_1-x_2)(x_1-x_3)}$$
$$+ \frac{y_2}{(x_2-x_0)(x_2-x_1)(x_2-x_3)} + \frac{y_3}{(x_3-x_0)(x_3-x_1)(x_3-x_2)}$$

and in general

$$(5)\qquad [x_0x_1\ldots x_n] = \frac{y_0}{(x_0-x_1)\ldots(x_0-x_n)}$$
$$+ \frac{y_1}{(x_1-x_0)(x_1-x_2)\ldots(x_1-x_n)} + \ldots$$
$$+ \frac{y_n}{(x_n-x_0)(x_n-x_1)\ldots(x_n-x_{n-1})}.$$

From the expression (5) we see at once the important fact that <u>the value of any divided difference is entirely independent of the order of the x's involved in the difference</u>.

Exercises

1. Find the fourth divided difference for the values

x :	0	1	2	3	4
y :	1	2	17	82	257

2. Find the fourth divided difference of $y = \sin x$, for the values $x_0 = 0^\circ$, $x_1 = 5^\circ$, $x_2 = 10^\circ$, $x_3 = 15^\circ$, $x_4 = 20^\circ$. (Take values of sin x to four places, and carry the calculations to five places.)

3. Show that

$$\text{(a)} \quad [x_0x_1] = \frac{\begin{vmatrix} 1 & y_0 \\ 1 & y_1 \end{vmatrix}}{\begin{vmatrix} 1 & x_0 \\ 1 & x_1 \end{vmatrix}} \qquad \text{(b)} \quad [x_0x_1x_2] = \frac{\begin{vmatrix} 1 & x_0 & y_0 \\ 1 & x_1 & y_1 \\ 1 & x_2 & y_2 \end{vmatrix}}{\begin{vmatrix} 1 & x_0 & x_0^2 \\ 1 & x_1 & x_1^2 \\ 1 & x_2 & x_2^2 \end{vmatrix}}$$

4. Show that

$$[x_0x_1\ldots x_n] = \frac{\begin{vmatrix} 1 & x_0 & \cdots & x_0^{n-1} & y_0 \\ 1 & x_1 & \cdots & x_1^{n-1} & y_1 \\ - & - & - & - & - \\ 1 & x_n & \cdots & x_n^{n-1} & y_n \end{vmatrix}}{\begin{vmatrix} 1 & x_0 & \cdots\cdots & x_0^n \\ 1 & x_1 & \cdots\cdots & x_1^n \\ - & - & - & - \\ 1 & x_n & \cdots\cdots & x_n^n \end{vmatrix}}$$

(Hint. Expand numerator determinant in terms of cofactors of the last column; show that the equation above reduces to equation (5) of this Article.)

5. Show that the nth divided difference of $y = x^n$ is unity, no matter what points $x_0, x_1, \ldots, x_n$ are chosen. (Hint. Set $y = x^n$ in Exercise 4 above.)

6. If y is a polynomial of degree n in x show that the nth divided differences are constant (i.e. independent of the x's) and that all higher order differences vanish.

7. If $y = x^n$ show that

$$\left[x_0x_1\right] = x_0^{n-1} + x_0^{n-2}x_1 + \ldots + x_0x_1^{n-2} + x_1^{n-1}.$$

8. If $y = x^5$ show that

$$\left[x_0x_1x_2\right] = x_0^3 + x_0^2\left[x_1 + x_2\right] + x_0\left[x_1^2 + x_1x_2 + x_2^2\right] + \left[x_1^3 + x_1^2x_2 + x_1x_2^2 + x_2^3\right].$$

9. If $y = (x - x_0)(x - x_1) \ldots (x - x_k)$, show that

$$\left[x_0x_1\ldots x_p\right] = 0, \qquad p \leqq k.$$

10. If $y = (x - x_0)(x - x_1) \ldots (x - x_k)$, show that

$$\left[x_0x_1\ldots x_p\right] = 0, \qquad p \geqq k + 2.$$

11. If $y = (x - x_0)(x - x_1) \ldots (x - x_k)$, show that

$$\left[x_0x_1\ldots x_{k+1}\right] = 1.$$

60. THE INTERPOLATING POLYNOMIAL IN TERMS OF DIVIDED DIFFERENCES

If we construct a set of divided differences for the values

$$x, x_0, x_1, x_2, \ldots, x_n$$

we have the equations

$$[x\ x_0] = \frac{y - y_0}{x - x_0}, \text{ whence } y = y_0 + [x\ x_0]\ (x - x_0)$$

and

$$|x\ x_0 x_1| = \frac{[x\ x_0] - [x_0 x_1]}{x - x_1}, \quad \text{whence}$$

$$[x\ x_0] = [x_0 x_1] + [x\ x_0 x_1]\ (x - x_1).$$

From the first equation on the right we eliminate $[x\ x_0]$ by substituting the valuo given for it by the second equation on the right. The result is

$$(1) \qquad y = y_0 + [x_0 x_1]\ (x - x_0) + [x\ x_0 x_1]\ (x - x_0)(x - x_1).$$

Now from

$$[x\ x_0 x_1 x_2] = \frac{[x\ x_0 x_1] - [x_0 x_1 x_2]}{x - x_2}$$

we have

$$[x\ x_0 x_1] = [x_0 x_1 x_2] + [x\ x_0 x_1 x_2]\ (x - x_2).$$

Substituting this value for the expression $[x\ x_0 x_1]$ in (1) we get

$$(2) \quad y = y_0 + [x_0 x_1]\ (x - x_0) + [x_0 x_1 x_2]\ (x - x_0)(x - x_1) + [x\ x_0 x_1 x_2]\ (x - x_0)(x - x_1)(x - x_2).$$

Proceeding step by step in this fashion we obtain the general formula (where $[x_0] = y_0$).

$$(3) \quad y(x) = \left[x_0\right] + \left[x_0x_1\right](x - x_0) + \left[x_0x_1x_2\right]$$

$$(x - x_0)(x - x_1)$$

$$+ \left[x_0x_1x_2x_3\right](x - x_0)(x - x_1)(x - x_2)$$

$$+ \cdots\cdots\cdots\cdots\cdots\cdots$$

$$+ \left[x_0x_1x_2\ldots x_n\right](x - x_0)(x - x_1)\ldots(x - x_{n-1})$$

$$+ \left[x\ x_0x_1\ldots x_n\right](x - x_0)(x - x_1)\ldots(x - x_n).$$

This formula is identically true for all values of x and for all functions y (which have definite values for the given x's). The importance of formula (3) justifies our giving it a careful examination. We note:

(a) All the divided differences involved, except the last one, are mere constants, since they do not contain the variable x.

(b) Each divided difference involves all the x's in the preceding divided difference, together with one new x.

(c) Each divided difference is multiplied by all the factors of the form $(x - x_i)$, such that x_i is contained in the <u>preceding</u> divided difference.

(d) The right-hand member of (3), excluding the last term, is a polynomial in x of degree not exceeding n. It we denote this polynomial by $P_n(x)$ we may write the identity (3) in the form

$$(4) \qquad y(x) = P_n(x)$$

$$+ \left[x\ x_0x_1\ldots x_n\right](x - x_0)(x - x_1)\ldots(x - x_n).$$

(e) Since (4) is an identity in x, we may set $x = x_0$, $x = x_1$, . . ., $x = x_n$ in succession and obtain the n+1 equations

$$y_k = P_n(x_k), \qquad k = 0, 1, \ldots, n.$$

It follows from this that $P_n(x)$ is the interpolating polynomial for $y = f(x)$ at the n+1 points $x_0, x_1, \ldots, x_n$.

(f) From equation (4) and the theorem of Article 23, we see that

$$(5) \qquad \left[x\, x_0 x_1 \ldots x_n\right] = \frac{f^{(n+1)}(s)}{(n+1)!}$$

where s lies between the greatest and least of the numbers $x, x_0, x_1, \ldots, x_n$.

(g) Using (5) and the fact that $y = f(x)$ we may write (3) in the form

$$\begin{aligned} f(x) = y_0 &+ \left[x_0x_1\right] (x - x_0) \\ &+ \left[x_0x_1x_2\right] (x - x_0)(x - x_1) \\ &+ \left[x_0x_1x_2x_3\right] (x - x_0)(x - x_1)(x - x_2) \\ &+ \ldots + \left[x_0x_1\ldots x_n\right] (x - x_0)(x - x_1)\ldots(x - x_{n-1}) \\ &+ \frac{f^{(n+1)}(s)}{(n+1)!} (x - x_0)(x - x_1)\ldots(x - x_n). \end{aligned}$$

This is the fundamental formula for polynomial interpolation in terms of divided differences.

Example 1. Obtain the interpolation formula for $y = f(x)$, given the points

x:	2	3	5	6.
y:	5	2	3	4

We first construct the table of divided differences

x	y	1st DD	2nd DD	3rd DD
2	<u>5</u>			
		<u>-3</u>		
3	2		<u>7/6</u>	
		$\frac{1}{2}$		<u>$-\frac{1}{4}$</u>
5	3		1/6	
		1		
6	4			

Then we substitute the numerical values of $[x_0]$, $[x_0x_1]$, $[x_0x_1x_2]$, etc., taken from the top diagonal of the above table into formula (6), which gives

$$f(x) = 5 - 3(x-2) + \tfrac{7}{6}(x-2)(x-3) - \tfrac{1}{4}(x -2(x-3)(x-5)$$

$$+ \frac{f^{(4)}(s)}{4!}(x-2)(x-3)(x-5)(x-6),$$

where s lies between the greatest and least of the values x, 2, 3, 5, 6. This is the required result. If desired, it may be arranged in increasing powers of x, but this is usually not necessary. By direct substitution of the values $x = 2, 3, 5, 6$, we may verify that the formula does indeed give the values 5, 2, 3, 4.

<u>Example 2</u>. Find the polynomial which gives the values

x:	3	5	8	11	15.
y:	6	-8	28	16	136

Proceeding as in the last example we have

x	y				
3	6				
		-7			
5	-8		18/5		
		11		-89/120	
8	25		-7/3		27/224
		-3		74/105	
11	16		33/7		
		30			
15	136				

and the required polynomial is

$$y = 6 - 7(x-3) + \frac{18}{5}(x-3)(x-5)$$

$$- \frac{89}{120}(x-3)(x-5)(x-8) + \frac{27}{224}(x-3)(x-5)(x-8)(x-11).$$

Example 3. Find a polynomial which coincides with $\sin x$ at $x = 21^\circ, 22^\circ, 24^\circ, 25^\circ$. From it calculate $\sin 23^\circ$.

Here the table of divided differences is

x	$\sin x^\circ$			
21	.35837			
		1624		
22	.37461		-6	
		1606		0,
24	.40674		-6	
		1588		
25	.42262			

and the required polynomial is

$$y = 0.35837 + 0.01624(x - 21) - 0.00006(x - 21)(x - 22).$$

Substituting $x = 23$ into this expression we get

$y = 0.39073$, which checks with $\sin 23^\circ$ to five places.

Exercises

1. Find the polynomial through the points

x:	1	3	4	6.
y:	4	10	19	79

2. Find the polynomial through the points

x:	0	1	3	4.
y:	13	11	-17	-55

3. Construct a table of divided differences for the function $y = x^2$ at the points 0, 1, 3, 5. 10.

4. In an experiment to determine the force required to tow a boat through water the following values were found

V (ft./sec.) :	30	50	70	80	90
F (lbs.) :	298.8	1328	3581	5317	7538

Find the force required for $V = 40$; 60; 75.

5. Given the following data

x	y
0.12	0.11943
0.19	0.18775
0.32	0.30951
0.41	0.38847
0.48	0.44624

find y a) when $x = 0.15$

b) when $x = 0.38$

6. Calculate (a) $\sin 20^\circ$, (b) $\sin 40^\circ$, (c) $\sin 50^\circ$,

from the values

φ	: 0°	30°	45°	60°	90° .
$\sin \varphi$	: 0	$\frac{1}{2}$	$\frac{1}{2}\sqrt{2}$	$\frac{1}{2}\sqrt{3}$	1

Calculate the remainder term and estimate the accuracy of your results.

The following pairs of values are known for a certain function $y = f(x)$.

x	y
0.10987	0.99396
0.22798	0.97367
0.30506	0.95233
0.41687	0.90897
0.51414	0.85771
0.59720	0.80210

7. Find y if x = 0.15932
8. Find y if x = 0.55636
9. Find y if x = 0.36162
10. Find x if y = 0.92866
11. Find x if y = 0.98877
12. From the values

x	y
-0.28213	1.0423
-0.08976	1.0041
0.03993	1.0008
0.13909	1.0098
0.21652	1.0243

find x when y = 1.0020. (See Article 22).

61. OTHER FORMS OF THE INTERPOLATING POLYNOMIAL

In the development of the interpolation formula (3) of the preceding article, we used the divided differences taken from the top diagonal of the difference table. This is only one of a number of ways in which the formula may be expressed. In fact if we examine the steps by which (3) was derived we see that we may use any selection of divided differences taken from the table provided only we do not violate the principles enunciated in (a), (b), and (c) of Article 60. These principles have been enunciated by Sheppard in the following Rule.

Sheppard's Rule

1. Start with any y_i.

2. The next term will be $[x_i x_j]\,(x-x_i)$. (Actually j is unrestricted but when the difference table has already been formed the only practical choice is either $j = i + 1$ or $j = i - 1$).

3. Each succeeding divided difference contains all the x's in the immediately preceding difference, together with one new x. (For practical purposes the new x must immediately precede or immediately follow the set of x's already used).

4. Each divided difference is multiplied by the product of factors of the form $(x - x_k)$ such that the product contains exactly the x's in the preceding difference.

Thus from the following table of divided differences

x_0	y_0				
		$[x_0x_1]$			
x_1	y_1		$[x_0x_1x_2]$		
		$[x_1x_2]$		$[x_0x_1x_2x_3]$	
x_2	y_2		$[x_1x_2x_3]$		$[x_0x_1x_2x_3x_4]$
		$[x_2x_3]$		$[x_1x_2x_3x_4]$	
x_3	y_3		$[x_2x_3x_4]$		
		$[x_3x_4]$			
x_4	y_4				

we might express the interpolating polynomial in any one of the forms:

(1) $$P(x) = y_0$$
$$+ [x_0x_1] (x - x_0) + [x_0x_1x_2] (x - x_0)(x - x_1)$$
$$+ [x_0x_1x_2x_3] (x - x_0)(x - x_1)(x - x_2)$$
$$+ [x_0x_1x_2x_3x_4] (x - x_0)(x - x_1)(x - x_2)(x - x_3).$$

This is the form given in (3), using the top diagonal.

(2) $$P(x) = y_4$$
$$+ [x_3x_4] (x - x_4) + [x_2x_3x_4] (x - x_4)(x - x_3)$$
$$+ [x_1x_2x_3x_4] (x - x_4)(x - x_3)(x - x_2)$$
$$+ [x_0x_1x_2x_3x_4] (x - x_4)(x - x_3)(x - x_2)(x - x_1).$$

This is formed with differences taken from the bottom diagonal.

(3) $$P(x) = y_2$$
$$+ [x_1x_2] (x - x_2) + [x_1x_2x_3] (x - x_1)(x - x_2)$$
$$+ [x_0x_1x_2x_3] (x - x_1)(x - x_2)(x - x_3)$$
$$+ [x_0x_1x_2x_3x_4] (x - x_0)(x - x_1)(x - x_2)(x - x_3).$$

This uses differences following a zigzag course through the table. All told there are 16 different formulas for this table, and in the general case for $n + 1$ points there are 2^n different formulas. But for a given set of values of x and y <u>all these different formulas express</u>

precisely the same interpolating polynomial.

As a matter of fact a still greater variety of formulas could be obtained by changing the order of the points denoted by x_0, x_1, ..., x_n, since nothing in the derivation of the formulas hinges on the order in which the actual values of x are taken. It is however usually convenient to arrange the x's in the order of increasing magnitude especially as no advantage results from changing the order.

On the other hand the different forms of the interpolating polynomial given in (1), (2), and (3) prove to be of real value since sometimes one, sometimes another, is found best adapted to a particular situation.

Exercises

1. Show that if the points x_0, x_1, . . ., are equally spaced, formula (1) above reduces to Newton's interpolation formula with forward differences, while (2) reduces to Newton's formula with backward differences.
2. Show that for equally spaced points, (3) reduces to Gauss's formula with backward differences.
3. Write out all the different forms of the interpolating polynomial that can be formed for the difference table above.
4. Applying the device used to obtain Stirling's formula, Article 50, derive the formula

$$y = y_0 + \frac{1}{2}\left(\left[x_1x_0\right] + \left[x_0x_{-1}\right]\right)(x - x_0)$$
$$+ \left[x_1x_0x_{-1}\right](x-x_0)\left(x-\frac{x_1+x_{-1}}{2}\right)$$
$$+ \frac{1}{2}\left(\left[x_2x_1x_0x_{-1}\right] + \left[x_1x_0x_{-1}x_{-2}\right]\right)(x-x_1)(x-x_0)(x-x_{-1})$$
$$+ \left[x_2x_1x_0x_{-1}x_{-2}\right](x-x_0)(x-x_1)(x-x_{-1})\left(x-\frac{x_2+x_{-2}}{2}\right)$$
$$+ \cdots .$$

Show that this reduces to Stirling's formula if the intervals are equal.

5. Obtain formulas which are generalizations of Everett's and Bessel's formulas.

6. Show that for equal intervals of length h

$$\left[x_0 x_1 \ . \ . \ . \ x_n\right] = \frac{\Delta^n y_o}{n! \ h^n} \ .$$

Chapter VIII
RECIPROCAL DIFFERENCES*

All the methods and formulas for interpolation so far developed have been based on the use of polynomials as approximating functions. In this chapter the approximating function is a rational fraction. Several fundamental differences from polynomial approximation are evident: the formulas are not linear in the undetermined coefficients; the values of the given function do not enter linearly; and the reasoning by which the error was determined for the polynomial case is no longer applicable.

62. APPROXIMATION BY MEANS OF RATIONAL FRACTIONS

Let $y = f(x)$ be a function of x for which the pairs of corresponding values (x_0, y_0), (x_1, y_1), (x_2, y_2), . . ., (x_n, y_n) are known. It is desired to obtain a rational fraction

$$(1) \qquad y = -\frac{a_0 + a_1x + a_2x^2 + \cdots}{b_0 + b_1x + b_2x^2 + \cdots}$$

in which the a's and b's are determined so that when $x = x_i$, $y = y_i$ for $i = 0, 1, \ldots, n$. Apparently the number of a's and b's required will be $n + 2$, since there are $n + 1$ conditions to be satisifed and one of the a's or b's can be chosen arbitrarily.

From (1) it follows that

$$a_0 + yb_0 + xa_1 + xyb_1 + x^2a_2 + x^2yb_2 + \cdots = 0$$

and since this is to be satisfied when $x = x_i$, $y = y_i$, $i = 0, 1, \ldots, n$, we have $n + 1$ additional equations

*See Appendix C.

$$a_0 + y_i b_0 + x_i a_1 + x_i y_i b_1 + x_i^2 a_2 + x_i^2 y_i b_2 + \cdots = 0$$

making $n + 2$ equations in all. The condition that non-zero values of the a's and b's exist satisfying these $n + 2$ equations is that the $(n + 2)$-rowed determinant should vanish, giving the equation

$$\begin{vmatrix} 1 & y_0 & x_0 & x_0 y_0 & x_0^2 & x_0^2 y_0 & \cdots \\ 1 & y_1 & x_1 & x_1 y_1 & x_1^2 & x_1^2 y_1 & \cdots \\ - & - & - & - & - & - & - \\ 1 & y_n & x_n & x_n y_n & x_n^2 & x_n^2 y_n & \cdots \\ 1 & y & x & xy & x^2 & x^2 y & \cdots \end{vmatrix} = 0$$

This is in effect the desired result. First of all, it is clear that this equation is satisfied whenever $x = x_i$, $y = y_i$, $i = 0, 1, \ldots, n$. Also, if the determinant be expanded in terms of cofactors of the last row and the resulting equation be solved for y, we see that y is expressed as the quotient of two polynomials.

<u>Example</u>. Given the set of values

x:	0	1	2	3
y:	4	2	4	7

express y as a rational fraction.

The determinant above becomes

$$\begin{vmatrix} 1 & 4 & 0 & 0 & 0 \\ 1 & 2 & 1 & 2 & 1 \\ 1 & 4 & 2 & 8 & 4 \\ 1 & 7 & 3 & 21 & 9 \\ 1 & y & x & xy & x^2 \end{vmatrix} = 0,$$

which when expanded by cofactors of the last row is

$$-4 + y + 4x + xy - 4x^2 = 0,$$

whence

$$y = 4\,\frac{1 - x + x^2}{1 + x}.$$

Exercises

Obtain rational fractions which have the assigned values:

1.

x:	0	1	2
y:	3	0	-1

2.

x:	-1	1	1	2	3
y:	0	10	0	-6	-8

3.

x:	0	1	2	3	4
y:	60	30	20	15	12

63. REDUCTION OF THE DETERMINANT

While the method outlined in Article 62 is theoretically satisfying, for practical computation it is necessary to devise some systematic procedure for reducing the order of the determinant. The method by which this reduction is achieved will be illustrated for the case of six points, $(x_0, y_0), \ldots, (x_5, y_5)$. The equation to be solved is then

$$(1)\quad \begin{vmatrix} 1 & y_0 & x_0 & x_0y_0 & x_0^2 & x_0^2y_0 & x_0^3 \\ 1 & y_1 & x_1 & x_1y_1 & x_1^2 & x_1^2y_1 & x_1^3 \\ - & - & - & - & - & - & - \\ 1 & y_5 & x_5 & x_5y_5 & x_5^2 & x_5^2y_5 & x_5^3 \\ 1 & y & x & xy & x^2 & x^2y & x^3 \end{vmatrix} = 0$$

We shall assume at first that the y's are all distinct, and proceed as follows:

1) Multiply 1st column by y_0 and subtract from 2nd column.
2) Multiply 3rd column by y_0 and subtract from 4th

column.

3) Multiply 5th column by y_0 and subtract from 6th column.

4) Multiply 5th column by x_0 and subtract from 7th column.

5) Multiply 3rd column by x_0 and subtract from 5th column.

6) Multiply 1st column by x_0 and subtract from 3rd column.

The first row is now 1, 0, 0, 0, 0, 0, 0, and we see that the original 7th order determinant is reduced to one of the 6th order. In this 6th order determinant:

7) Divide the 1st row by $y_1 - y_0$,

8) Divide the 2nd row by $y_2 - y_0$,

- - - - - - - - - - - - - - - - - -

9) Divide the last row by $y - y_0$.

10) Let

$$\frac{x - x_0}{y - y_0} = \rho_1(xx_0), \quad \frac{x_i - x_0}{y_i - y_0} = \rho_1(x_ix_0), \text{ etc.}$$

The result may now be written

$$(2) \quad \begin{vmatrix} 1 & \rho_1(x_1x_0) & x_1 & x_1\,\rho_1(x_1x_0) & x_1^2 & x_1^2\,\rho_1(x_1x_0) \\ 1 & \rho_1(x_2x_0) & x_2 & x_2\,\rho_1(x_2x_0) & x_2^2 & x_2^2\,\rho_1(x_2x_0) \\ - & - & - & - & - & - \\ 1 & \rho_1(x\,x_0) & x & x\,\rho_1(x\,x_0) & x^2 & x^2\,\rho_1(x\,x_0) \end{vmatrix} = 0$$

The determinant in (2) is seen to be of exactly the same type as that in (1) except that $\rho_1(x_jx_0)$ replaces y and the first row and last column are missing.

Let us now treat (2) in the same manner as we did (1). In the steps corresponding to 7), 8), 9), the divisors will be

$$\rho_1(x_2x_0) - \rho_1(x_1x_0), \; . \; . \; ., \; \rho_1(x\ x_0) - \rho_1(x_1x_0),$$

and we assume for the moment that these do not vanish. The final result is in the form

$$(3) \qquad \begin{vmatrix} 1 & u_2 & x_2 & x_2u_2 & x_2^2 \\ 1 & u_3 & x_3 & x_3u_3 & x_3^2 \\ - & - & - & - & - \\ 1 & u_x & x & xu_x & x^2 \end{vmatrix} = 0.$$

where for brevity we have set

$$u_x = \frac{x - x_1}{\rho_1(x\ x_0) - \rho_1(x_1x_0)}.$$

Now the quantity $\rho_1(x_ix_j)$ is obviously symmetrical in its subscripts so that $\rho_1(x_jx_i) = \rho_1(x_ix_j)$. But this is clearly not true of the expression u, in which for example the subscripts 0 and 1 are not interchangeable. It may, however, be verified by elementary algebra that the following identities hold:

$$\frac{x - x_1}{\dfrac{x - x_0}{v - v_0} - \dfrac{x_1 - x_0}{v_1 - v_0}} + v_0 \equiv \frac{x - x_0}{\dfrac{x - x_1}{v - v_1} - \dfrac{x_1 - x_0}{v_1 - v_0}} + v_1$$

$$\equiv \frac{x_1 - x_0}{\dfrac{x_1 - x}{v_1 - v} - \dfrac{x - x_0}{v - v_0}} + v.$$

Since symmetry is an essential desideratum in the sequel, we proceed to modify (3) still further by the following steps:

11) Multiply the 1st column by y_0 and add to the 2nd column.

12) Multiply the 3rd column by y_0 and add to the 4th column.

If we define the quantity $\rho_2(x\ x_1x_0)$ by the formula

$$\rho_2(x\ x_1x_0) = \frac{x - x_1}{\rho_1(x\ x_0) - \rho_1(x_1x_0)} + y_0 \tag{4}$$

equation (3) now becomes

$$\begin{vmatrix} 1 & \rho_2(x_2x_1x_0) & x_2 & x_2\,\rho_2(x_2x_1x_0) & x_2^2 \\ 1 & \rho_2(x_3x_1x_0) & x_3 & x_3\,\rho_2(x_3x_1x_0) & x_3^2 \\ - & - & - & - & - \\ 1 & \rho_2(x\ x_1x_0) & x & x\ \rho_2(x\ x_1x_0) & x^2 \end{vmatrix} = 0. \tag{5}$$

This determinant is again of the same type as that in (1) except that $\rho_2(x\ x_1x_0)$ replaces y and the first two rows and last two columns are missing. Applying to (5) the steps corresponding to 1), 2), . . ., 12) and noting that in 11) and 12) the multiplier is now $\rho_1(x_1x_0)$, we get

$$\begin{vmatrix} 1 & \rho_3(x_3x_2x_1x_0) & x_3 & x_3\ \rho_3(x_3x_2x_1x_0) \\ - & - & - & - \\ 1 & \rho_3(x\ x_2x_1x_0) & x & x\ \rho_3(x\ x_2x_1x_0) \end{vmatrix} = 0,$$

where

$$\rho_3(x\ x_2x_1x_0) = \frac{x - x_2}{\rho_2(x\ x_1x_0) - \rho_2(x_2x_1x_0)} + \rho_1(x_1x_0).$$

Again we easily show that $\rho_3(x_3x_2x_1x_0)$ is symmetrical in all its arguments. After sufficient repetitions of the process of reduction we arrive at the two-rowed determinant

$$\begin{vmatrix} 1 & \rho_5(x_5x_4x_3x_2x_1x_0) \\ 1 & \rho_5(x\ x_4x_3x_2x_1x_0) \end{vmatrix} = 0,$$

whence

(6) $$\rho_5(x\ x_4x_3x_2x_1x_0) = \rho_5(x_5x_4x_3x_2x_1x_0).$$

From the definitions of the ρ's it follows that

$$y = y_0 + \frac{x - x_0}{\rho_1(x\ x_0)}$$

$$\rho_1(x\ x_0) = \rho_1(x_1x_0) + \frac{x - x_1}{\rho_2(x\ x_1x_0) - y_0}$$

$$\rho_2(x\ x_1x_0) = \rho_2(x_2x_1x_0) + \frac{x - x_2}{\rho_3(x\ x_2x_1x_0) - \rho_1(x_1x_0)}$$

etc.

By eliminating in succession $\rho_1(x\ x_0)$, $\rho_2(x\ x_1x_0)$, etc., and finally using equation (6) to eliminate $\rho_5(x\ x_4x_3x_2x_1x_0)$, we derive the expression for y in the form of a continued fraction,

(7) $$y = y_0 + \cfrac{x - x_0}{\rho_1(x_1x_0) + \cfrac{x - x_1}{\rho_2(x_2x_1x_0) - y_0 + \cfrac{x - x_2}{\rho_3(x_3x_2x_1x_0) - \rho_1(x_1x_0)}}}$$

$$+ \ \cdots$$

terminating with

$$\frac{x - x_4}{\rho_5(x_5x_4x_3x_2x_1x_0) - \rho_3(x_3x_2x_1x_0)}$$

The procedure illustrated here for the case of six points can easily be extended to the general case.

64. RECIPROCAL DIFFERENCES

The quantities

$$\rho_1(x_1x_0) = \frac{x_1 - x_0}{y_1 - y_0},$$

$$\rho_2(x_2x_1x_0) = \frac{x_2 - x_1}{\rho_1(x_2x_0) - \rho_1(x_1x_0)} + y_0$$

- -

$$\rho_{n+1}(x_{n+1}\ldots x_0) = \frac{x_{n+1} - x_n}{\rho_n(x_{n+1}\ldots x_0) - \rho_n(x_n\ldots x_0)}$$

$$+ \rho_{n-1}(x_{n-1}\ldots x_0)$$

introduced in Article 63 are known as "reciprocal differences." As already indicated, the reciprocal difference $\rho_n(x_nx_{n-1}\ldots x_0)$ is symmetrical in all its arguments. This fact makes it possible to calculate the differences entering equation (7) of Article 63 by means of a simple table of differences, set up in the usual manner

x_0	y_0			
		$\rho_1(x_1x_0)$		
x_1	y_1		$\rho_2(x_2x_1x_0)$	
		$\rho_1(x_2x_1)$		$\rho_3(x_3x_2x_1x_0)$
x_2	y_2		$\rho_2(x_3x_2x_1)$	
		$\rho_1(x_3x_2)$		$\rho_3(x_4x_3x_2x_1)$
x_3	y_3		$\rho_2(x_4x_3x_2)$	
		$\rho_1(x_4x_3)$		
x_4	y_4			

For example in deriving formula (7) we obtained $\rho_3(x_3x_2x_1x_0)$ by the equation

$$\rho_3(x_3x_2x_1x_0) = \frac{x_3 - x_2}{\rho_2(x_3x_1x_0) - \rho_2(x_2x_1x_0)} + \rho_1(x_1x_0)$$

while in the table $\rho_3(x_3x_2x_1x_0)$ is found from the equation

$$\rho_3(x_3x_2x_1x_0) = \frac{x_3 - x_0}{\rho_2(x_3x_2x_1) - \rho_2(x_2x_1x_0)} + \rho_1(x_2x_1),$$

but the property of symmetry enables us to show that the results are the same.

A marked advantage in the arrangement of the reciprocal difference table as shown above is that <u>any</u> downward diagonal of differences can be used to construct the partial fraction. For example we have

$$y = y_1 + \cfrac{x - x_1}{\rho_1(x_2x_1) + \cfrac{x - x_2}{\rho_3(x_3x_2x_1) - y_1 + \cfrac{x - x_3}{\text{etc.}}}}$$

<u>Example 1</u>. Construct the table of reciprocal differences for the set of values

x:	0	1	2	3	4	5	6
y:	1	$\frac{1}{2}$	$\frac{1}{5}$	$\frac{1}{10}$	$\frac{1}{17}$	$\frac{1}{26}$	$\frac{1}{37}$

and obtain a rational fraction which takes on these values.

The table of differences is

0	1				
		-2			
1	1/2		-1		
		-10/3		0	
2	1/5		-1/10		0
		-50/5		40	
3	1/10		-1/25		0
		-170/7		140	
4	1/17		-1/46		0
		-442/9		324	
5	1/26		-1/73		
		-962/11			
6	1/37				

The partial fraction formed according to (7) from the differences in the first diagonal is

$$y = 1 + \cfrac{x - 0}{-2 + \cfrac{x - 1}{-2 + \cfrac{x - 2}{2 + \cfrac{x - 3}{1}}}},$$

and this, after reduction to a simple fraction, turns out to be

$$y = \frac{1}{1 + x^2}$$

The continued fraction formed from the reciprocal differences in the second diagonal is likewise

$$y = \frac{1}{2} + \cfrac{x - 1}{-10/3 + \cfrac{x - 2}{- 6/10 + \cfrac{x - 3}{130/3 + \cfrac{x - 4}{1/10}}}}$$

which also in this case reduces to

$$y = \frac{1}{1 + x^2} .$$

The method of reciprocal differences may be used for interpolation in the vicinity of points where the function becomes infinite and where in consequence polynomial interpolation is unsuitable.

Example 2. From the tabulated values of csc x for $x = 1^\circ, 2^\circ, 3^\circ, 4^\circ$ calculate csc $1^\circ 30'$.

Here the table of reciprocal differences will be

x	csc x	ρ_1	ρ_2	ρ_3
1°	57.298677			
2°	28.653706	-0.034910142		
3°	19.107321	-0.10475169	0.017457	
4°	14.335588	-0.20956747	0.026225	342.05

The continued fraction formed for the top diagonal and with $x = 1.5$ is

$$y = 57.298677 + \cfrac{0.5}{-0.034910142 + \cfrac{-0.5}{-57.281220 + \cfrac{-1.5}{342.08}}}$$

which reduces to

$$y = 38.201548$$

The value given in the tables for csc $1^\circ 30'$ is 38.201547.

Instead of calculating the interpolated value of y directly from the continued fraction as was done above, another method is to employ the successive convergents of the continued fraction

$$y = a_0 + \cfrac{x - x_0}{a_1 + \cfrac{x - x_1}{a_2 + \cfrac{x - x_2}{a_3 + \cdots}}}$$

where for brevity we let

$$a_n = \rho_n(x_n x_{n-1} \dots x_0) - \rho_{n-2}(x_{n-2} \dots x_0).$$

If $z_n = p_n/q_n$ denotes the nth convergent of the foregoing fraction, we have by the theory of continued fractions

$$\begin{array}{lll} p_0 = 1 \quad , & q_0 = 0 \quad , & \\ p_1 = a_0 p_0 \quad , & q_1 = 1 \quad , & z_1 = p_1/q_1 \\ p_2 = a_1 p_1 + (x - x_0)p_0 \quad , & q_2 = a_1 q_1 \quad , & z_2 = p_2/q_2 \\ p_3 = a_2 p_2 + (x - x_1)p_1 \quad , & q_3 = a_2 q_2 + (x - x_1)q_1 \quad , & z_3 = p_3/q_3 \end{array}$$

- -

$$p_{n+1} = a_n p_n + (x - x_{n-1})p_{n-1}\ , \quad q_{n+1} = a_n q_n + (x - x_{n-1})q_{n-1}, \quad z_{n+1} = p_{n+1}/q_{n+1}$$

Accordingly, the computation may be arranged in the tabular form

n	a_n	$x - x_n$	p_n	q_n	z_n
0	a_0	$x - x_0$	1	0	
1	a_1	$x - x_1$	p_1	1	z_1
2	a_2	$x - x_2$	p_2	q_2	z_2
3	a_3	$x - x_3$	p_3	q_3	z_3
- -	- - -	- - - - -	- - -	- - -	- -

The values of the a's are taken from the table of reciprocal differences, the values of $x - x_1$ for the desired x are entered, the p's and q's can then be obtained by the recurrence relations above, and finally the successive convergents are found by $z_n = p_n/q_n$. For Example 2 the computation is shown below.

a_n	$x - x_n$	p_n	q_n	z_n
57.298677	0.5	1	0	
-0.034910142	-0.5	57.298677	1	57.298677
-57.281220	-1.5	-1.50030495	-0.034910142	42.976192
342.08		57.289957	1.4996955	38.201059
		19599.9989	513.06820	38.201547

While this procedure effects no saving of labor, it has the virtue of indicating whether or not the process is converging toward a fixed value.

Exercises

Form the table of reciprocal differences for:

1. $\frac{1}{1 + x}$ for $x = 0, 1, 2, 3$.

2. $\frac{1 + x^2}{1 + x}$ for $x = 0, 2, 3, 4$.

3. x^3 for $x = 0, 1, \ldots, 6$.

4. $\frac{1 - x^2}{1 + x^2}$ for $x = 1, 2, 3, \ldots, 6$.

5. tan x for $x = 85^\circ, 86^\circ, 87^\circ, 88^\circ, 89^\circ$.
   Use the continued fraction to find tan $85^\circ 20'$.

6. sec x for $x = 85^\circ, 86^\circ, 87^\circ, 88^\circ, 89^\circ$.
   Use the continued fraction to find sec $85^\circ 40'$.

7. Show how upward diagonals in the difference table may be used to form the continued fraction just as downward diagonals have been used.

8. In exercises 5 and 6 find tan $88^{\circ}30'$ and sec $88^{\circ}20'$.

65. FURTHER PROPERTIES OF RECIPROCAL DIFFERENCES

Reciprocal differences may be expressed as the quotient of two determinants, as shown by the examples

$$\rho_4(x_0x_1x_2x_3x_4) = \frac{\begin{vmatrix} 1 & y_i & x_i & x_iy_i & x_i^2y_i \end{vmatrix}}{\begin{vmatrix} 1 & y_i & x_i & x_iy_i & x_i^2 \end{vmatrix}},$$

$$\rho_5(x_0x_1x_2x_3x_4x_5) = \frac{\begin{vmatrix} 1 & y_i & x_i & x_iy_i & x_i^2 & x_i^3 \end{vmatrix}}{\begin{vmatrix} 1 & y_i & x_i & x_iy_i & x_i^2 & x_i^2y_i \end{vmatrix}},$$

where each determinant is indicated by a typical row. To show this, we follow the steps of the reduction of the determinant

$$\begin{vmatrix} 1 & y_0 & x_0 & x_0y_0 & x_0^2 & x_0^2y_0 \\ 1 & y_1 & x_1 & x_1y_1 & x_1^2 & x_1^2y_1 \\ - & - & - & - & - & - \\ 1 & y_4 & x_4 & x_4y_4 & x_4^2 & x_4^2y_4 \\ 1 & y & x & x\,y & x^2 & x^2y \end{vmatrix} = 0$$

Let A_0 and B_0 denote respectively the cofactors of the last and next to the last elements in the bottom row of this determinant, A_1, B_1, A_2, B_2, . . ., being similarly defined for the first reduced determinant, second reduced determinant, etc. We note that all columns of A_i and B_i are alike except the last column. In the reduction of the determinant A_i and B_i remain unchanged by the combination of columns except in those cases where the last column of A_i is combined with the last column of B_i. Both determinants are equally affected by the divisions. Hence, if D_i denotes the product of the divisors at any step we have after the first step

$$A_1 = A_0/D_1,$$

$$B_1 = (B_0 - y_0A_0)/D_1.$$

Similarly after the second step

$$A_2 = A_1/D_2,$$

$$B_2 = B_1/D_2 + A_2y_0.$$

Hence, after two steps we have

$$B_2/A_2 = B_0/A_0$$

After two more steps we have

$$B_4/A_4 = B_2/A_2 = B_0/A_0.$$

But the determinant has now been reduced to

$$\begin{vmatrix} 1 & \rho_4(x_0x_1x_2x_3x_4) \\ 1 & \rho_4(x_0x_1x_2x_3x\ \ ) \end{vmatrix} = 0$$

from which we see that $A_4 = 1$, $B_4 = -\rho_4(x_0x_1x_2x_3x_4)$, whence

$$\rho_4(x_0x_1x_2x_3x_4) = -B_0/A_0.$$

The proof for any case of even order is similar. In the case of odd order the first step gives

$$B_1/A_1 = B_0/A_0$$

the next two give

$$B_3/A_3 = B_1/A_1$$

and the argument proceeds as before.

It is convenient to introduce the term "order" of a rational fraction, defined as follows. Let $y = N(x)/D(x)$ be a rational fraction which is irreducible (i.e., having no common polynomial factor in numerator and denominator). Let the actual degree of N(x) be m, and the actual degree of D(x) be n. The order k of the rational fraction is defined to be

$$k = 2n \quad \text{if} \quad m \leqq n, \qquad k = 2m - 1 \quad \text{if} \quad m > n.$$

If k is odd

$$y = \frac{a_0 + a_1 x + \ldots + a_m x^m}{b_0 + b_1 x + \ldots + b_{m-1} x^{m-1}} \qquad a_m \neq 0.$$

If k is even

$$y = \frac{a_0 + a_1 x + \ldots + a_n x^n}{b_0 + b_1 x + \ldots + b_n x^n} \qquad b_n \neq 0.$$

In either case there are $k + 2$ constants, one of which is arbitrary, since numerator and denominator may be divided by any non-zero coefficient.

<u>Theorem 1</u>. <u>If</u> $(x_0, y_0)(x_1, y_1)\ldots(x_k, y_k)$ <u>are</u> $k + 1$ <u>points with distinct</u> x's, <u>there cannot exist two distinct irreducible rational fractions</u> $y = N_1(x)/D_1(x)$ <u>and</u> $y = N_2(x)/D_2(x)$ <u>of order</u> $\leq k$ <u>both of which are satisfied by the given</u> $k + 1$ <u>pairs of values</u> (x_i, y_i).

For if both fractions are satisfied then the equation

$$N_1(x) \cdot D_2(x) = N_2(x) \cdot D_1(x)$$

is true for $k + 1$ distinct values of x. But each member of this equation is a polynomial of degree not exceeding k, and hence, the two members must be identically equal and must contain identical linear factors (real or imaginary). Moreover, since $N_1(x)$ and $D_1(x)$ have no common factors, all the linear factors of $N_1(x)$ must be in $N_2(x)$ and all the factors of $D_1(x)$ must be in $D_2(x)$. Likewise, all linear factors of $N_2(x)$ must be in $N_1(x)$ and all linear factors of $D_2(x)$ must be in $D_1(x)$. Hence, $N_1(x)/D_1(x)$ and $N_2(x)/D_2(x)$ are identical fractions.

Theorem 2. *If $y = R_k(x)$ is an irreducible rational fraction of order k, the reciprocal differences of order k are constant.*

For using $x_1, x_2, \ldots, x_{k+1}$ we get by the reduction of Article 64

$$\rho_k(x\, x_k \ldots x_1) = \rho_k(x_{k+1}\, x_k \ldots x_1),$$

while using $x_1, x_2, \ldots, x_k, x_0$ we have

$$\rho_k(x\, x_k \ldots x_1) = \rho_k(x_0 x_k \ldots x_1)$$

$$= \rho_k(x_k \ldots x_0)$$

because of symmetry. Hence

$$\rho_k(x_k \ldots x_0) = \rho_k(x_{k+1} \ldots x_1).$$

In this way, step by step, we show that all differences of order k are equal.

Theorem 3. *If in a table of reciprocal differences formed for n pairs (x_i, y_i) the kth differences are constant, $k < n - 1$, then all n values satisfy $y = R_k(x)$ where $R_k(x)$ is a rational fraction of order k.*

This may be seen at once from the determinant form. If for example ρ_4 is a constant then ρ_5 is infinite so that the determinant

$$\begin{vmatrix} 1 & y & x & xy & x^2 & x^2y \end{vmatrix} = 0$$

for every set of 5 distinct pairs (x_i, y_i). Since ρ_4 is by assumption finite the determinant

$$\begin{vmatrix} 1 & y & x & xy & x^2 \end{vmatrix}$$

does not vanish, and hence the above equation defines a rational fraction of order 4.

66. EXCEPTIONAL CASES

Up to this point the discussion of reciprocal differences and approximation by rational fractions is based on the assumption that no hitch occurs in carrying out the indicated operations. As a matter of fact several kinds of difficulties may arise.

1) The determinant which we set up to obtain the rational fraction may vanish identically. It can be shown that this will occur if, and only if, the $k + 1$ pairs of given values satisfy a rational fraction $y = R(x)$ of order not exceeding $k - 2$.

The nature of the general proof will be indicated by a particular example. Suppose that there are five given points and that the determinant is

$$\begin{vmatrix} 1 & y & x & xy & x^2 & x^2y \end{vmatrix} = 0 \tag{1}$$

Now suppose that all five points satisfy a relationship of the type

$$a + by + cx + dxy = 0 \qquad d \neq 0$$

Then they also satisfy

$$ax + bxy + cx^2 + dx^2y = 0.$$

Using these two relations we can combine columns in (1) so as to have two columns with only zeros in the first five rows. It is now apparent that all the cofactors of the last row vanish and the determinant vanishes identically.

Conversely suppose that the determinant vanishes identically and consequently all cofactors of the last row vanish. In particular

$$\begin{vmatrix} 1 & y_i & x_i & x_iy_i & x_i^2 \end{vmatrix} = 0 \tag{2}$$

and

$$\begin{vmatrix} 1 & y_i & x_i & x_iy_i & x_i^2y_i \end{vmatrix} = 0 \tag{3}$$

where $i = 0, 1, \ldots, 3$ in the first, second, . . ., fourth rows respectively. Let the cofactors of the elements of some row, say the first row, of (2) be A, B, C, D, E, and for (3) let them be A', B', C', D', E. Note that E is the same in both. Then

$$A + By + Cx + Dxy + Ex^2 = 0,$$

$$A' + B'y + C'x + D'xy + Ex^2y = 0,$$

will both be satisfied by the 5 pairs (x_i, y_i). The equation in x obtained by eliminating y is of the form

$$E^2x^4 + \text{lower powers} = 0.$$

Since this equation of fourth degree is satisfied by five distinct values of x it vanishes identically and in particular $E = 0$. In the same way we show that all the cofactors of the last column in (2) and (3) must vanish. The first four columns are therefore linearly dependent so that there exists a relation

$$a + by + cx + dxy = 0$$

which is satisfied by all five points.

2) The determinant set up to obtain the rational fraction may factor into two rational factors. Since the determinant is linear in y the factorization must be of the form

$$P(x)\ Q(x, y) = 0 \tag{4}$$

where without loss of generality we may assume that $Q(x, y)$ is irreducible. Let the degree of $P(x)$ be p and let $Q(x, y) = 0$ define a rational fraction $y = R(x)$ of order q. If $k + 1$ points were used in forming the original determinant we must have $q + 2p \leqq k$. Also since all these points satisfy (4), and at most p of them can satisfy $P(x) = 0$ the remainder, $k + 1 - p$, must satisfy $y = R(x)$. Now $q + 1$ points are required to determine $R(x)$, so that the equation $y = R(x)$ is satisfied by $k + 1 - p - (q + 1) = k - p - q$ points in excess of the number required to determine $R(x)$. This excess we shall for brevity designate as S "surplus" points. Hence

$$S = k - p - q$$

and by the inequality above

$$S \geqq p.$$

Hence if the determinant factors into

$$P(x)\ Q(x, y) = 0$$

where $p(x)$ is of degree p the rational fraction $y = R(x)$ obtained from $Q(x, y) = 0$ must be satisfied by at least p surplus points.

Obviously in such a case the attempt to represent all $k + 1$ given values by means of a rational fraction of order k is doomed to failure.

Example 1. Find the rational fraction for the values

x:	0	1	2	3
y:	1	1	2	3

Here the determinant is

$$\begin{vmatrix} 1 & 1 & 0 & 0 & 0 \\ 1 & 1 & 1 & 1 & 1 \\ 1 & 2 & 2 & 4 & 4 \\ 1 & 3 & 3 & 9 & 9 \\ 1 & y & x & xy & x^2 \end{vmatrix} = 0$$

whence

$$xy = x^2$$

This equation is satisfied by all four values but the equation $y = x$ obtained from it is not. For this problem $k = 3$, $q = 1$, $p = 1$, $S = 1$.

3) The process of reducing the determinant or of constructing a table of divided differences may be halted by the presence of zero divisors.

When this occurs two or more divided differences of the same order must be equal, say

$$\rho_3(x_1x_2x_3x_4) = \rho_3(x_0x_1x_2x_3).$$

By Theorem 3 of Article 65 this implies the existence of a rational fraction $y = R_3(x)$ of order three satisfied by all five pairs (x_0, y_0), . . ., $(x_4\ y_4)$.

4) The continued fraction by which the desired representation of y is expressed may fail to give back the original values for which it was constructed.

It can, however, be shown that such a situation

will not occur for a fraction terminating with ρ_k if no two reciprocal differences of the same order less than k are equal.

The foregoing discussion of exceptional cases is by no means exhaustive but only points out the most common difficulties. If we use the term "degenerate set" to describe a set of $n + 1$ points which satisfy a rational fraction of order less than n we may summarize our results as follows:

Theorem. *If a set of $n + 1$ points is not degenerate and contains no degenerate subset there exists an irreducible rational fraction $y = R_n(x)$ of order n which is satisfied by the given points. Aside from a common constant factor in numerator and denominator the fraction is unique.*

The conditions in the theorem are sufficient but not necessary, as shown by the example

$$y = \frac{1 + x^2}{1 + x}$$

which is satisfied by the values

x:	0	1	2	3
y:	1	1	5/3	5/2

These values contain the degenerate subset (0, 1), (1, 1).

Exercises

1. Given the points

x:	0	1	2	3
y:	12	0	-4	-6

find the corresponding rational fraction.

2. Show that the set in Exercise 1 is degenerate.

3. Show that the set

x:	0	1	2	3	4
y:	12	0	-4	-6	6

satisfies the equation

$$(x - 4)(12 - y - 12x + xy) = 0.$$

4. Show that there is no irreducible rational fraction of order four or less which is satisfied by the points in Exercise 3.
5. Show that the set

x:	0	1	2	3	4	5
y:	12	0	-4	-6	6	4

contains a degenerate subset, but that the entire set is satisfied by a rational fraction of order five. Obtain this fraction.

Ans. $$y = \frac{1680 - 2478x + 897x^2 - 99x^3}{140 + 24x - 17x^2}$$

Chapter IX

POLYNOMIAL APPROXIMATION BY LEAST SQUARES*

In earlier chapters the approximating polynomial was determined in such a way as to coincide with the value of the given function at definite points. In certain types of problems this may be quite undesirable, particularly if the given values have been experimentally obtained and are therefore subject to random errors. We do not wish to incorporate these errors in the approximating function but seek rather to find an approximation which reflects the general trend of the given function without reproducing local fluctuations. The method of Least Squares is designed to achieve this end.

67. LEAST SQUARES

Let there be given $n + 1$ pairs of values (x_0, y_0), (x_1, y_1), . . ., (x_n, y_n) of a function $y = f(x)$, and a polynomial

(1) $$y_m(x) = a_0 + a_1 x + \cdots + a_m x^m$$

of degree m, where $m < n$. We propose to determine the unknown coefficients $a_0, a_1, \ldots, a_m$ in such a way that the sum of the squares of the differences between f(x) and $y_m(x)$ is a minimum. We have then to obtain the minimum of the expression

(2) $$S = \sum_{j=0}^{n} \left[f(x_j) - a_0 - a_1 x_j - \cdots - a_m x_j^m \right]^2 ,$$

in which S is a function of the $m + 1$ independent variables $a_0, a_1, \ldots, a_m$. We proceed in the usual

*See Appendix C.

manner to obtain the partial derivatives of S with respect to a_0, a_1, . . ., a_m and to equate these derivatives to zero. In this way we arrive at the $m + 1$ equations

$$-2 \sum_{j=0}^{n} \left[f(x_j) - a_0 - a_1 x_j - \cdots - a_m x_j^m \right] x_j^k = 0$$

$$(k = 0, 1, \ldots, m)$$

which are rewritten in the more convenient form

$$a_0 \sum_{j=0}^{n} x_j^k + a_1 \sum_{j=0}^{n} x_j^{k+1} + \cdots + a_m \sum_{j=0}^{n} x_j^{k+m}$$
$$= \sum_{j=0}^{n} y_j x_j^k$$

for $k = 0, 1, \ldots, m$. Here $f(x_j)$ has been replaced by y_j. Using for brevity the notation

$$s_k = \sum_{j=0}^{n} x_j^k, \qquad v_k = \sum_{j=0}^{n} y_j x_j^k$$

these equations become

$$
\begin{aligned}
s_0 a_0 + s_1 a_1 + \cdots + s_m a_m &= v_0, \\
s_1 a_0 + s_2 a_1 + \cdots + s_{m+1} a_m &= v_1, \\
\cdots\cdots\cdots\cdots\cdots & \\
s_m a_0 + s_{m+1} a_1 + \cdots + s_{2m} a_m &= v_m.
\end{aligned} \tag{3}
$$

It will be shown in the next article that the determinant of the coefficients in these equations does not vanish so that the equations always have a unique solution for a_0, a_1, . . ., a_m. Furthermore when the values so obtained for the a's are substituted into (1) it will be shown that the value of S in (2) is actually a minimum.

The numerical calculation involved in carrying out the foregoing procedure can be systematically arranged in tabular form as shown in the following scheme for the case $m = 2$, $n = 3$.

x^0	x	x^2	x^3	x^4	y	xy	x^2y
1	x_0	x_0^2	x_0^3	x_0^4	y_0	x_0y_0	$x_0^2y_0$
1	x_1	x_1^2	x_1^3	x_1^4	y_1	x_1y_1	$x_1^2y_1$
1	x_2	x_2^2	x_2^3	x_2^4	y_2	x_2y_2	$x_2^2y_2$
1	x_3	x_3^2	x_3^3	x_3^4	y_3	x_3y_3	$x_3^2y_3$
s_0	s_1	s_2	s_3	s_4	v_0	v_1	v_2

$$s_0a_0 + s_1a_1 + s_2a_2 = v_0$$

$$s_1a_0 + s_2a_1 + s_3a_2 = v_1$$

$$s_2a_0 + s_3a_1 + s_4a_2 = v_2$$

The given values of x and y are entered in the columns headed x and y respectively, the powers and products are computed, and each column is added to obtain the s's and v's. The equations for the a's can then be set up. Since evidently the determinant of these equations is always symmetrical the method of Article 6 may be used for their solution. When the a's have been found the approximating polynomial is

$$y(x) = a_0 + a_1x + a_2x^2.$$

Example 1. Fit a polynomial of degree 2 to the following data:

x	y
-3	-0.71
-2	-0.01
-1	0.51
0	0.82
1	0.88
2	0.81
3	0.49

The calculation is shown below.

x^0	x	x^2	x^3	x^4	y	xy	x^2y
1	-3	9	-27	81	-.71	2.13	-6.39
1	-2	4	-8	16	-.01	.02	-.04
1	-1	1	-1	1	.51	-.51	.51
1	0	0	0	0	.82	0	0
1	1	1	1	1	.88	.88	.88
1	2	4	8	16	.81	1.62	3.24
1	3	9	27	81	.49	1.47	4.41
7	0	28	0	196	2.79	5.61	2.61

The equations (3) become

$$7a_0 \qquad + \quad 28a_2 = 2.79$$

$$28a_1 \qquad = 5.61$$

$$28a_0 \qquad 196a_2 = 2.61$$

whence $a_0 = 0.806$, $a_1 = 0.200$, $a_2 = -0.102$ to 3 decimal places and the desired polynomial is

$$y = 0.806 + 0.200x - 0.102x^2.$$

Example 2. Fit a polynomial of degree 2 to the data

x	y
0.78	2.50
1.56	1.20
2.34	1.12
3.12	2.25
3.81	4.28

The work, carried to 3 places of decimals, appears as follows:

x^0	x	x^2	x^3	x^4	y	xy	x^2y
1	0.78	0.608	0.475	0.370	2.50	1.950	1.520
1	1.56	2.434	3.796	5.922	1.20	1.872	2.921
1	2.34	5.476	12.813	29.982	1.12	2.621	6.133
1	3.12	9.734	30.371	94.759	2.25	7.020	21.902
1	3.81	14.516	55.306	210.717	4.28	16.307	62.128
5	11.61	32.768	102.761	341.750	11.350	29.770	94.604

Here we set up and solve the equations by the method of Article 6.

5	11.61	32.768	11.350
11.61	32.768	102.761	29.770
32.768	102.761	341.750	94.604
5	2.322	6.554	2.270
11.61	5.810	4.590	0.588
32.768	26.689	4.486	1.009
5.045	-4.043	1.009	

Hence $y = 5.045 - 4.043x + 1.009x^2$.

It is of interest to compare the original values of y with those obtained by substitution in the approximating formula. We find

x	Original y	Approximate y	Difference
0.78	2.50	2.505	-0.005
1.56	1.20	1.194	+0.006
2.34	1.12	1.110	+0.010
3.12	2.25	2.252	-0.002
3.81	4.28	4.288	-0.008

If it is assumed that the original values of y were subject to random errors, it is not unreasonable to suppose that the values given by the approximation are better than the original values, since the approximation tends to smooth out local irregularities.

Quite frequently it is desirable to represent a set of experimental data by means of an equation of the type

$$S = qt^p \tag{4}$$

in which q and p are constants to be determined in a manner appropriate to the data. One way of doing this is to set

$$\begin{aligned} y &= \log S \\ a_0 &= \log q \\ a_1 &= p \\ x &= \log t. \end{aligned}$$

Then the above equation is equivalent to

$$y = a_0 + a_1 x$$

which is a polynomial of degree 1, and where the coefficients a_0 and a_1 can now be found by least squares, as shown above. The logarithms involved may be taken to the base 10.

Another problem of frequent occurrence is to represent a set of data by the formula

$$S = Ae^{ct}, \tag{5}$$

in which A and c are to be determined. This can be reduced as before to

$$y = a_0 + a_1 x$$

by the substitutions

$$y = \ln S, \ a_0 = \ln A, \quad a_1 = c, \quad x = t.$$

Natural logarithms are used here, though of course common logs could be employed by defining $a_1 = c \log_{10} e$.

Example 3. Fit a formula of type (4) to the data

t	S
2.2	65
2.7	60
3.5	53
4.1	50

The calculation in compact form is shown below.

	$x = \log t$	x^2	$y = \log S$	xy
1	0.3424	0.1172	1.8129	0.6207
1	0.4314	0.1861	1.7782	0.7671
1	0.5441	0.2960	1.7243	0.9382
1	0.6128	0.3755	1.6990	1.0411
4	1.9307	0.9748	7.0144	3.3671

$$4\,a_0 + 1.9307a_1 = 7.0144$$

$$1.9307a_0 + 0.9748a_1 = 3.3671$$

4	0.4827	1.7536
1.9307	0.0428	-0.434

$\log q = 1.963 \qquad -0.434$

$q = 91.9 \qquad\qquad p = -0.434$

$$S = 91.9t^{-0.434}$$

Exercises

Obtain by least squares the best approximation to the given data by a polynomial of degree 1.

1.

x:	1	2	3	4	5	6	7	8	9	10
y:	4	3	6	7	11	11	13	18	18	20

2.

x:	0	0.1	0.2	0.3	0.4	0.5	0.6	0.7
y:	3.02	2.81	2.57	2.39	2.18	1.99	1.81	1.85

3. Represent the given data by a formula of type (4).

t:	1	2	3	4	5
s:	7.1	27.8	62.1	110	161

4. Represent the given data by a formula type (5).

t:	0	2	4	6	8	10	12
s:	1280	635	324	162	76	43	19

5. Represent the given data for y by a quadratic function of x.

x:	7	8	9	10	11	12	13
y:	7.4	8.4	9.1	9.4	9.5	9.5	9.4

68. FURTHER INVESTIGATION OF LEAST SQUARES

In Article 67, we left unanswered the question whether the equations could always be solved and also whether the solution, if obtainable, always yields a minimum for S. These points will now be considered.

First of all we must establish some important properties of a special type of determinant. Let $x_0, x_1, \ldots, x_n$ be $n + 1$ distinct values of the real variable x, and let s_k be defined by the equation

$$s_k = x_0^k + x_1^k + x_2^k + \ldots + x_n^k. \tag{1}$$

By D_m we shall denote the m-rowed symmetrical determinant.

$$D_m = \begin{vmatrix} s_0 & s_1 & s_2 & \cdots & s_{m-1} \\ s_1 & s_2 & s_3 & \cdots & s_m \\ s_2 & s_3 & s_4 & \cdots & s_{m+1} \\ \cdots & \cdots & \cdots & \cdots & \cdots \\ s_{m-1} & s_m & s_{m+1} & \cdots & s_{2m-2} \end{vmatrix} \tag{2}$$

By $y_m(x)$ let us denote the polynomial of degree m defined by the equation

$$(3) \qquad y_m(x) = \begin{vmatrix} s_0 & s_1 & \cdots & s_m \\ s_1 & s_2 & \cdots & s_{m+1} \\ \cdots & \cdots & \cdots & \cdots \\ s_{m-1} & s_m & \cdots & s_{2m-1} \\ 1 & x & \cdots & x^m \end{vmatrix}$$

Let us multiply (3) by x^k, where $k < m$, putting the factor x^k in the bottom row of the determinant. We then sum over all values $x_0, x_1, \ldots, x_n$, and obtain

$$(4) \qquad \sum_{i=0}^{n} x_i^k \, y_m(x_i) = 0 \qquad (k < m)$$

because the bottom row becomes $s_k \quad s_{k+1} \ \ldots \ s_{k+m}$ and is identical with one of the preceding rows for $k < m$. On the other hand if we multiply (3) by x^m and sum we find that

$$(5) \qquad \sum_{i=0}^{n} x_i^m \, y_m(x_i) = D_{m+1},$$

since the bottom row becomes $s_m \quad s_{m+1} \ \ldots \ s_{2m}$, and the resulting determinant is now D_{m+1} as shown by (2).

Now let us multiply (3) by $y_m(x)$, putting the factor $y_m(x)$ in the last row of the determinant. We next sum over all values $x_0, x_1, \ldots, x_n$, and obtain

$$(6) \qquad \sum_{i=0}^{n} y_m^2 \, (x_i) = D_m \, D_{m+1}.$$

For the bottom row of the determinant becomes $0 \quad 0 \ \ldots \ 0 \quad D_{m+1}$ by virtue of equations (4) and (5), and evidently the cofactor of the element in the lower right-hand corner is D_m.

It will be convenient to use the symbol S_m defined as follows:

$$S_m = \sum_{i=0}^{n} y_m{}^2(x_i), \tag{7}$$

from which we have at once

$$S_m = D_m\, D_{m+1}.$$

Now S_m, being a sum of squares, can vanish only if $y_m(x_0)$, $y_m(x_1)$, . . ., $y_m(x_n)$ are all zero, that is, only if the polynomial y_m vanishes at all the points x_0, x_1, . . ., x_n. For $n \geq m$ this can occur only if $y_m(x)$ is identically zero. It is evident however that $y_1(x)$ is not identically zero, since S_0, the coefficient of x, is equal to $n + 1$. Therefore, $S_1 > 0$. Now $D_1 = |S_0| = n + 1$, and setting $m = 1$ in (8) we have

$$S_1 = D_1\, D_2$$

whence $D_2 > 0$. Again setting $m = 2$ in (8) we have

$$S_2 = D_2\, D_3.$$

Since the coefficient of x^2 in $y_2(x)$ is D_2 which does not vanish, we see that S_2 is positive, hence $D_3 > 0$. We may continue this reasoning step by step until the m in (8) becomes equal to n. After that point the reasoning fails because the polynomial $y_{n+1}(x)$ could vanish at the points x_0, x_1, . . ., x_n without vanishing identically. The conclusion of the whole matter is that

$$D_1,\ D_2,\ \ldots,\ D_{n+1}$$

are all positive.

We now take up the method of least squares. We are to choose the coefficients a_k in the polynomial

$$a_0 + a_1 x + \dots + a_m x^m$$

so as to make the expression

$$S = \sum_{i=0}^{n} \left[f(x_i) - a_0 - a_1 x_i - \dots - a_m x_i^m \right]^2$$

(where $m \leqq n$) a minimum.

A necessary condition is that the $m + 1$ partial derivatives $\frac{\partial S}{\partial a_k}$, $(k = 0, 1, \dots, m)$, all vanish. We then have

$$\frac{\partial S}{\partial a_k} = - \sum_{i=0}^{n} 2x_i^k \left[f(x_i) - a_0 - a_1 x_i - \dots - a_m x_i^m \right] = 0,$$

from which we get the $m + 1$ equations

$$(9) \qquad s_k a_0 + s_{k+1} a_1 + \dots + s_{k+m} a_m = \sum_{i=0}^{n} x_i^k f(x_i),$$

$$(k = 0, 1, 2, \dots, m)$$

for the determination of the $m + 1$ coefficients. The determinant of the equations (9) is precisely D_{m+1}, which has been shown to be positive, so that equations (9) always have a unique solution. The equations (9) are easily seen to be identical with the equations that were used for the determination of the coefficients in the examples of the preceding article.

It remains to inquire whether this method actually yields the minimum value for the sum of the squares. It is shown in advanced calculus that a sufficient condition for a minimum of a function S of several variables $a_0, a_1, \ldots, a_m$ is that

$$\frac{\partial S}{\partial a_0}, \quad \frac{\partial S}{\partial a_1}, \quad \ldots, \quad \frac{\partial S}{\partial a_m}$$

all vanish, and in addition, for the values of the a's which make the above partial derivatives vanish, the following determinants

$$\left|\frac{\partial^2 S}{\partial a_0^2}\right|, \quad \begin{vmatrix} \frac{\partial^2 S}{\partial a_0^2} & \frac{\partial^2 S}{\partial a_0\,\partial a_1} \\ \frac{\partial^2 S}{\partial a_0\,\partial a_1} & \frac{\partial^2 S}{\partial a_1^2} \end{vmatrix}$$

etc., up to

$$\begin{vmatrix} \frac{\partial^2 S}{\partial a_0^2} & & \frac{\partial^2 S}{\partial a_0\,\partial a_m} \\ \cdot & \cdots & \cdot \\ \frac{\partial^2 S}{\partial a_m\,\partial a_0} & & \frac{\partial^2 S}{\partial a_m^2} \end{vmatrix}$$

are all positive. We easily find that $\dfrac{\partial^2 S}{\partial a_j\,\partial a_k} = 2\sum_{i=0}^{n} x_i^{j+k} = 2s_{j+k}$, so that the determinants above are respectively equal to $2D_1$, 2^2D_2, . . ., $2^{m+1}D_{m+1}$, and are therefore positive, since the D's have been shown to be positive.

Consequently since equations (9) have but one

solution and this solution gives a minimum, we have found the absolute minimum for the sum of the squares of the errors.

69. LEAST SQUARES FOR INTEGRALS

The two preceding Articles dealt with the case where the coefficients of a polynomial were chosen so as to make the sum of the squares of the error a minimum. We now take up the case where the object is to make the integral of the square of the error a minimum, that is, we are to choose coefficients a_k in the polynomial $a_0 + a_1x + \dots + a_mx^m$ so that

$$\int_a^b \left[f(x) - a_0 - a_1x - \dots - a_mx^m\right]^2 dx$$

has the smallest possible value. This problem can be treated almost word for word as in the case of least squares for sums, the only difference being that sums, wherever they occur, are replaced by integrals. However, it turns out that a better method is available, which we shall now explain.

First of all it is convenient to change the variable so that the interval of integration $a \leqq x \leqq b$ is replaced by the standard interval 0 to 1. This can always be accomplished by the substitution $x = (b - a)\,x' + a$ We shall assume henceforth that this change has already been made, and that the interval is 0 to 1.

In the second place it will be shown in the next article that there exists a sequence of polynomials

$$P_0(x) = 1$$
$$P_1(x) = 1 - 2x$$
$$P_2(x) = 1 - 6x + 6x^2$$
$$P_3(x) = 1 - 12x + 30x^2 - 20x^3$$
$$\dots\dots\dots\dots\dots\dots$$
$$P_m(x) = \sum_{k=0}^{m} (-1)^k \binom{m}{k} \binom{m+k}{k} x^k$$

which possess the useful property that

$$(1) \qquad \int_0^1 P_m(x)\, P_n(x)\, dx = \begin{cases} 0 \text{ if } m \neq n \\ \dfrac{1}{2m+1} \text{ if } m = n \end{cases}$$

Instead of trying to determine the coefficients in $a_0 + a_1 x + \dots + a_m x^m$ we shall determine the coefficients $c_0, c_1, \dots, c_m$ in the expression

$$c_0 P_0(x) + c_1 P_1(x) + \dots + c_m P_m(x)$$

(which is evidently a polynomial of degree m) in such a manner that

$$(2) \quad I = \int_0^1 \left[f(x) - c_0 P_0(x) - c_1 P_1(x) - \dots - c_m P_m(x) \right]^2 dx$$

shall be a minimum. Differentiating with respect to c_k we have

$$\frac{\partial I}{\partial c_k} =$$

$$\int_0^1 -2P_k(x) \left[f(x) - c_0 P_0(x) - c_1 P_1(x) - \dots - c_m P_m(x) \right] dx$$

$$= -2 \int_0^1 P_k(x)\, f(x)\, dx + \frac{2c_k}{2k + 1}$$

when use is made of formulas (1). Setting these partial derivatives equal to zero and solving for the c_k we have

$$(3) \qquad c_k = (2k + 1) \int_0^1 P_k(x)\, f(x)\, dx,$$

$$(k = 0, 1, \ldots, m).$$

The second order partial derivatives are found to be

$$\frac{\partial^2 I}{\partial c_i\, \partial c_k} = \begin{cases} 0 & \text{if } i \neq k \\ \dfrac{2}{2k + 1} & \text{if } i = k. \end{cases}$$

All the determinants $\left| \dfrac{\partial^2 I}{\partial c_i\, \partial c_k} \right|$ are therefore positive and the values of the c's determined in (3) consequently give an actual minimum. The approximating polynomial is

$$(4) \qquad \sum_{k=0}^{m} \left[(2k + 1) \int_0^1 P_k(x)\, f(x)\, dx \right] P_k(x),$$

as we find by putting the values of the c's given by (3) back into the original polynomial.

70. ORTHOGONAL POLYNOMIALS

We now take up the derivation of the set of polynomials $P_m(x)$ introduced in Article 69, which have the important property expressed in equation (1) Article 69.

Let us try to construct a polynomial $P_m(x)$ of degree m,

$$(1) \qquad P_m(x) = 1 + a_1 x + a_2 x^2 + \ldots + a_m x^m$$

such that

$$(2) \qquad \int_0^1 x^s P_m(x)\,dx = 0 \qquad \text{for} \quad s = 0, 1, \ldots, m - 1.$$

We multiply (1) by $x^s\,dx$, integrate from $x = 0$ to $x = 1$, and use (2) to obtain m linear equations

$$(3) \qquad \frac{1}{s + 1} + \frac{a_1}{s + 2} + \ldots + \frac{a_m}{s + m + 1} = 0$$

$$(s = 0, 1, \ldots, m - 1)$$

for the determination of the m coefficients a_1, a_2, ..., a_m. We could solve these equations by determinants, but an easier way is the following:

If we reduce the left-hand member of (3) to a simple fraction, we get

$$(4) \qquad \frac{1}{s + 1} + \frac{a_1}{s + 2} + \ldots + \frac{a_m}{s + m + 1} = \frac{Q(s)}{(s + m + 1)^{(m + 1)}}$$

in which the numerator $Q(s)$ is a polynomial in s of degree not higher than m. The expression $(s + m + 1)^{(m + 1)}$ in the denominator denotes the factorial $(s + m + 1)(s + m)(s + m - 1) \ldots$ to $(m + 1)$ factors. Since by (3) the polynomial $Q(s)$ must vanish for the m values of s, $s = 0, 1, \ldots, m - 1$, it follows that $Q(s) = Cs^{(m)}$ in which C is a constant. If we multiply (4) by $s + 1$ and set $s = -1$ in the resulting equation we find that $C = (-1)^m$. Therefore

$$(5)\qquad \frac{1}{s+1}+\frac{a_1}{s+2}+\ .\ .\ .\ +\frac{a_m}{s+m+1} = \frac{(-1)^m\, s^{(m)}}{(s+m+1)^{(m+1)}}$$

If we multiply (5) by $(s+m+1)^{(m+1)}$ and then set $s = -(k+1)$, where k is an integer not greater than m, we have

$$(-1)^k\, k!\,(m-k)!\; a_k = (-1)^m\,(-k-1)^{(m)} = (m+k)^{(m)}.$$

Solving this equation for a_k we have

$$(6)\qquad a_k = (-1)^k \frac{(m+k)^{(m)}}{(m-k)!\;k!} = (-1)^k \binom{m}{k}\binom{m+k}{k}\;.$$

We now put the values of the a's obtained from (6) into equation (1) and express the polynomial $P_m(x)$ in the form

$$(7)\qquad P_m(x) = \sum_{k=0}^{m} (-1)^k \binom{m}{k}\binom{m+k}{k}\; x^k.$$

This is the desired polynomial which satisfies the equations (2). There is one polynomial $P_m(x)$ for each integral value of m. The first nine are explicitly given as follows:

$$\begin{aligned}
&P_0(x) = 1 \\
&P_1(x) = 1 - 2x \\
&P_2(x) = 1 - 6x + 6x^2 \\
&P_3(x) = 1 - 12x + 30x^2 - 20x^3 \\
&P_4(x) = 1 - 20x + 90x^2 - 140x^3 + 70x^4 \\
(8) \quad &P_5(x) = 1 - 30x + 210^2 - 560x^3 + 630x^4 - 252x^5 \\
&P_6(x) = 1 - 42x + 420x^2 - 1680x^3 + 3150x^4 - 2772x^5 \\
&\qquad + 924x^6 \\
&P_7(x) = 1 - 56x + 756x^2 - 4200x^3 + 11550x^4 - 16632x^5 \\
&\qquad + 12012x^6 - 3432x^7 \\
&P_8(x) = 1 - 72x + 1260x^2 - 9240x^3 + 34650x^4 - 72072x^5 \\
&\qquad + 84084x^6 - 51480x^7 + 12870x^8.
\end{aligned}$$

If we multiply equation (1) by $P_n(x)$ and integrate from 0 to 1 we have

$$(9) \quad \int_0^1 P_m(x)\, P_n(x)\, dx = \sum_{k=0}^{m} a_k \int_0^1 x^k\, P_n(x)\, dx.$$

If $m < n$ all the integrals on the right vanish because of (2) so that

$$(10) \quad \int_0^1 P_m(x)\, P_n(x)\, dx = 0, \qquad (m \neq n).$$

(If $m > n$, we interchange m and n in (9) and get the same result.)

If m = n all the integrals on the right in (9) vanish except the last and (9) becomes

$$\int_0^1 P_m^2(x)\,dx = a_m \int_0^1 x^m P_m(x)\,dx.$$

The value of the integral on the right is found from (5) by setting $s = m$, and proves to be

$$\int_0^1 x^m P_m(x)\,dx = \frac{(-1)^m\, m^{(m)}}{(2m+1)^{(m+1)}}.$$

The value of a_m obtained from (6) is

$$a_m = (-1)^m \binom{2m}{m}.$$

Using these expressions we obtain finally

$$\int_0^1 P_m^2(x)\,dx = \frac{1}{2m+1}. \tag{11}$$

Equations (10) and (11) express very important properties of the polynomials $P_m(x)$, and as will appear later are of great value in solving problems of approximation by Least Squares. Functions having the property (10) are called "orthogonal functions." The orthogonal polynomials $P_m(x)$ are known as Legendre Polynomials* in honor of A. M. Legendre.

From equations (2) it is apparent that if Q(x) denotes any polynomial of degree less than m then

*Strictly speaking the term "Legendre polynomials" refers to orthogonal polynomials formed for the interval -1 to 1. A simple change of variable will convert our $P_m(x)$ to the customary form.

$$\int_0^1 Q(x)\, P_m(x)\, dx = 0. \tag{12}$$

Since $P_m(x)$ is of degree m the equation $P_m(x) = 0$ has m roots, real or complex, each root counted a number of times equal to its multiplicity. Let $r_1, r_2, \ldots, r_k$, $k \leqq m$, denote the odd order real roots that lie in the interval $0 < x < 1$. Then clearly the product

$$(x - r_1)(x - r_2) \ldots (x - r_k)\, P_m(x)$$

does not change sign in the interval $0 < x < 1$, since all its zeros in this interval are of even order. Consequently the integral

$$\int_0^1 (x - r_1)(x - r_2) \ldots (x - r_k)\, P_m(x)\, dx$$

does not vanish. Then we must have $k = m$, since otherwise the integral would vanish by (12). It follows that all the roots of $P_m(x) = 0$ are real, distinct, and lie in the interval $0 < x < 1$.

Five place tables of $P_m(x)$ up to $m = 5$ are given in Table V.

Exercises

1. By putting $m = 0, 1, 2, 3, 4$, in turn in formula (7) obtain the expressions for $P_0(x)$, $P_1(x)$, $P_2(x)$, $P_3(x)$, $P_4(x)$ given in (8).
2. By direct integration using the formulas in (8), verify (10) and (11) for $m = 0, 1, 2, 3$, $n = 0, 1, 2, 3$.
3. Show that $P_m(1) = (-1)^m$.
4. Show that $P_m(1 - x) = (-1)^m P_m(x)$.
5. If m is odd show that $P_m(\frac{1}{2}) = 0$.

6. Plot the curves for $P_0(x)$, $P_1(x)$, $P_2(x)$, $P_3(x)$, $P_4(x)$, $P_5(x)$ in the interval 0 to 1 using Table V.
7. From the results of 6 verify the statement that the roots are real, distinct, and lie in the interval $0 < x < 1$.

71. USE OF ORTHOGONAL POLYNOMIALS

We return now to the actual problem of approximating by Least Squares as discussed in Article 69. The procedure may be most readily explained by use of an example.

Example. Approximate the function $f(x) = \frac{4}{2 + x}$ in the interval $2 < x < 6$ by a polynomial of degree 5, using least squares.

First of all, the variable is changed so that the interval becomes $0 < s < 1$ by the substitution $x = 4s + 2$, which transforms the given function to

$$\frac{1}{1 + s}$$

Using the explicit values of the $P_m(x)$ as given in (8) Article 70, we have

$$c_0 = \int_0^1 \frac{1}{1 + s}\, ds = \log 2,$$

$$c_1 = 3 \int_0^1 \frac{1}{1 + s}(1 - 2s)\, ds = 9 \log 2 - 6,$$

$$c_2 = 5 \int_0^1 \frac{1}{1 + s}(1 - 6s + 6s^2)\, ds = 65 \log 2 - 45.$$

Similarly

$$c_3 = 441 \log 2 - 917/3,$$

$$c_4 = 2889 \log 2 - 2002.5$$

$$c_5 = 18513 \log 2 - 384967/30.$$

Using log 2 = 0.69314718055 (the large number of places is necessary to obtain accuracy in the c's) we have

$$c_0 = 0.693147, \qquad c_3 = 0.011240,$$
$$c_1 = 0.238325, \qquad c_4 = 0.002205,$$
$$c_2 = 0.054567, \qquad c_5 = 0.000421.$$

The desired polynomial, in terms of the variable s, is therefore

$$Q(s) = 0.693147 + 0.238325\ P_1(s) + 0.054567\ P_2(s)$$
$$+ 0.011240\ P_3(s) + 0.002205\ P_4(s) + 0.000421\ P_5(s).$$

It is of interest to compare the values given by this approximating polynomial with the true values of the function $\frac{1}{s+1}$. Using the Table we find the following values.

s	Value of Polynomial	Value of $\frac{1}{s+1}$	Value of $\frac{1}{s+1}$ - Polynomial
0	.99991	1.00000	+ .00009
0.1	.90913	.90909	- .00004
0.2	.83331	.83333	+ .00002
0.3	.76921	.76923	+ .00002
0.4	.71430	.71429	- .00001
0.5	.66669	.66667	- .00002
0.6	.62500	.62500	.00000
0.7	.58821	.58824	+ .00003
0.8	.55555	.55556	+ .00001
0.9	.52634	.52632	- .00002
1.0	.49993	.50000	+ .00007

Exercises

1. Using the method outlined in this Article, find the best approximation to x^4 in the interval $0 < x < 1$ by a polynomial of degree 3.

2. Find the best approximation to x^5 by a polynomial of degree 4.

3. Find the best approximation to $\frac{1}{x + 2}$ in the interval $0 < x < 1$ by a polynomial of degree 4. Plot the function and the approximating polynomial to a large scale on the same graph.

4. Find the best approximation to $\frac{1}{x + 1}$ in the interval $1 < x < 2$ by a polynomial of degree 3.

5. Find the best approximation to $\sin \frac{\pi x}{2}$ in the interval $0 < x < 1$ by a polynomial of degree 3.

6. $\sqrt{1 + x}$ $\quad 0 < x < 1$ $\quad$ degree 4.

7. $\log (4 + x)$ $\quad 0 < x < 1$ $\quad$ degree 3.

8. $\cos x$ $\quad 0 < x < 1$ $\quad$ degree 4.

9. $\frac{1 - x}{1 + x}$ $\quad 0 < x < 1$ $\quad$ degree 5.

10. If m is a positive integer show that

$$x^m = \frac{(m!)^2}{(2m + 1)!} \sum_{k=0}^{m} (-1)^k (2k + 1) \binom{2m+1}{m-k} P_k(x).$$

72. ORTHOGONAL POLYNOMIALS. EQUALLY SPACED POINTS

The method of Least Squares as outlined in Articles 67, 68 becomes exceedingly laborious if the degree of the approximating polynomial is greater than 3 or 4. It is therefore highly important to develop methods that require less numerical labor. In the case of equally spaced points this can be accomplished by the use of orthogonal polynomials entirely analogous to the Legendre polynomials used in Articles 69, 70, 71 for integrals.

Let there be given a set of $n + 1$ equally spaced points $x_0, x_1, \ldots, x_n$ with interval h. Since the change of variable, $x' = \frac{x - x_0}{h}$, carries these points over into 0, 1, 2, . . ., n, we shall assume that such a transformation has been made, and that the points are always taken as 0, 1, 2, . . ., n.

Let us try to construct a polynomial $P_{m,n}(x)$ of degree m

$$(1) \qquad P_{m,n}(x) = 1 + b_1x + b_2x^{(2)} + . . . + b_mx^{(m)}$$

such that

$$(2) \qquad \sum_{x=0}^{n} (x + s)^{(s)} P_{m,n}(x) = 0$$

for $s = 0, 1, 2, . . ., m - 1$. If we multiply (1) by $(x + s)^{(s)}$ we have

$$(x + s)^{(s)} P_{m,n}(x) = (x + s)^{(s)} + b_1(x + s)^{(s+1)} + . . . + b_m(x + s)^{(s+m)}.$$

and when this is summed from $x = 0$ to $x = n$ we have

$$(3) \qquad \sum_{x=0}^{n} (x + s)^{(s)} P_{m,n}(x) = \frac{(n + s + 1)^{(s+1)}}{s + 1} + b_1 \frac{(n + s + 1)^{(s+2)}}{s + 2} + . . . + b_m \frac{(n + s + 1)^{(s+m+1)}}{s + m + 1} = 0$$

if $s = 0, 1, 2, . . ., m - 1$. From this equation we remove the factor $(n + s + 1)^{(s+1)}$ leaving

$$\frac{1}{s + 1} + \frac{nb_1}{s + 2} + \frac{n^{(2)}b_2}{s + 3} + . . . + \frac{n^{(m)}b_m}{s + m + 1} = 0$$

for $s = 0, 1, . . ., m - 1$. If we set $n^{(k)}b_k = a_k$, $k = 1, 2, . . ., m$, we get

$$\frac{1}{s+1} + \frac{a_1}{s+2} + \ . \ . \ . + \frac{a_m}{s+m+1} = 0.$$

These equations are identical with (3) Article 70, and their solutions are found from (6) Article 70 to be

$$a_k = (-1)^k \binom{m}{k} \binom{m+k}{k},$$

so that

$$\text{(4)} \qquad b_k = (-1)^k \binom{m}{k} \binom{m+k}{k} \frac{1}{n^{(k)}}.$$

When these values are put in (1) the final expression for the polynomial $P_{m,n}(x)$ proves to be

$$P_{m,n}(x) = \sum_{k=0}^{m} (-1)^k \binom{m}{k} \binom{m+k}{k} \frac{x^{(k)}}{n^{(k)}} .$$

The first six of these polynomials, explicitly written out, are

$$P_{0,n}(x) = 1,$$

$$P_{1,n}(x) = 1 - 2\frac{x}{n},$$

$$P_{2,n}(x) = 1 - 6\frac{x}{n} + 6\frac{x(x-1)}{n(n-1)},$$

$$\text{(6)} \qquad P_{3,n}(x) = 1 - 12\frac{x}{n} + 30\frac{x(x-1)}{n(n-1)} - 20\frac{x(x-1)(x-2)}{n(n-1)(n-2)},$$

$$P_{4,n}(x) = 1 - 20\frac{x}{n} + 90\frac{x(x-1)}{n(n-1)} - 140\frac{x(x-1)(x-2)}{n(n-1)(n-2)}$$

$$+ 70\frac{x(x-1)(x-2)(x-3)}{n(n-1)(n-2)(n-3)},$$

$$P_{5,n}(x) = 1 - 30\frac{x}{n} + 210\frac{x(x - 1)}{n(n - 1)} - 560\frac{x(x - 1)(x - 2)}{n(n - 2)(n - 2)}$$

$$+ 630\frac{x(x - 1)(x - 2)(x - 3)}{n(n - 1)(n - 2)(n - 3)}$$

$$- 252\frac{x(x - 1)(x - 2)(x - 3)(x - 4)}{n(n - 1)(n - 2)(n - 3)(n - 4)}.$$

Any polynomial $Q(x)$ of degree q can be expressed as follows:

$$Q(x) = \sum_{s=0}^{q} A_s(x + s)^{(s)}$$

We multiply this equation by $P_{m,n}(s)$ and sum with respect to x from $x = 0$ to $x = n$, obtaining

$$\sum_{x=0}^{n} Q(x)\ P_{m,n}(x) = 0 \tag{7}$$

when $q < m$, as we easily see from equations (2). In particular

$$\sum_{x=0}^{n} P_{m,n}(x)\ P_{k,n}(x) = 0 \tag{8}$$

provided $m \neq k$. This is the orthogonal property.

Using methods similar to those employed for the derivation of (11), Article 70, we may also show that

$$\sum_{x=0}^{n} P^2_{m,n}(x) = \frac{(n + m + 1)(m + n)^{(m)}}{(2m + 1)(n)^{(m)}} \tag{9}$$

Table VI gives the values of $P_{m,n}(x)$ for $n = 5, 6, 7, \ldots, 20$, for $m = 1, 2, 3, 4, 5$, and for $x = 0, 1, 2, \ldots, n$. A word of explanation of these tables is desirable, and for this purpose we give here the table for $n = 10$ (11 points).

x	$P_{1,10}(x)$	$P_{2,10}(x)$	$P_{3,10}(x)$	$P_{4,10}(x)$	$P_{5,10}(x)$
0	5	15	30	6	3
1	4	6	-6	-6	-6
2	3	-1	-22	-6	-1
3	2	-6	-23	-1	4
4	1	-9	-14	4	4
5	0	-10	0	6	0
6	-1	-9	14	4	-4
7	-2	-6	23	-1	-4
8	-3	-1	22	-6	1
9	-4	6	6	-6	6
10	-5	15	-30	6	-3
$S_{m,10}$	110	858	4290	286	156

First of all since $P_{0,n}(x) = 1$ for all values of n and x, the value of $P_{0,n}(x)$ is not given in the tables. Secondly, the values in the table are not the values of $P_{m,n}(x)$ as defined above but are the numerators of fractions whose denominators are the entries at the head of the column, corresponding to $x = 0$. Thus $P_{3,10}(2) = -\frac{22}{30}$, $P_{4,10}(7) = -\frac{1}{6}$, $P_{5,10}(4) = +\frac{4}{3}$, etc. Finally, at the bottom of the column is given $S_{m,n}$, the sum of the squares of the numbers in that column. The value of

$$\sum_{x=0}^{n} P_{m,n}^{2}(x)$$

is given by $S_{m,n}$ divided by the square of the entry at the head of the column. Thus, e.g.

$$\sum_{x=0}^{10} P_{3,10}^{2}(x) = \frac{4290}{30^2} .$$

The tables are arranged in this manner to avoid the necessity of printing fractions or decimals.

As a matter of fact this arrangement of the tables makes them also more convenient in applications, for we can completely ignore the fact that the tabular value is not actually $P_{m,n}(x)$, because the factor introduced to give integral values cancels out in the formula for the approximating polynomial. The approximating polynomial is

$$Q_m(x) = \sum_{k=0}^{m} \left[\frac{\sum_{x=0}^{n} P_{k,n}(x)\, f(x)}{\sum_{x=0}^{n} P_{k,n}^2(x)} \right] P_{k,n}(x) \tag{10}$$

and we readily see from this expression that $P_{k,n}(x)$ may be multiplied by any non-zero factor independent of x without affecting the value of $Q_m(x)$. It is convenient to employ the notation

$$c_k = \sum_{x=0}^{n} P_{k,n}(x)\, f(x),$$

$$S_k = \sum_{x=0}^{n} P_{k,n}^2(x),$$

so that (10) becomes

$$Q_m(x) = \sum_{k=0}^{m} \frac{c_k}{S_k} P_{k,n}(x). \tag{11}$$

The magnitude of the error may be estimated by means of the expression

$$S = \sum_{x=0}^{n} \left[f(x) - Q_m(x) \right]^2,$$

which, by use of (11) and the orthogonal property of the

P's, can be expressed as follows:

$$S = \sum_{x=0}^{n} f^2(x) - \sum_{k=0}^{m} \frac{c_k^2}{S_k}. \tag{12}$$

This quantity S gives the sum of the squares of the deviations of the approximating polynomial Q(x) from the given function f(x) at the $n + 1$ points 0, 1, . . ., n.

Example 1. The numerical integration of a certain differential equation gave the following values of x and y:

x	y
0	1.300
0.3	1.245
0.6	1.095
0.9	0.855
1.2	0.514
1.5	0.037
1.8	-0.600
2.1	-1.295
2.4	-1.767
2.7	-1.914

Let us obtain by least squares a fifth degree polynomial which approximately represents y in this interval.

The calculation may be conveniently arranged as shown in the table below

s	x	y		$P_0(s)$	$P_1(s)$	$P_2(s)$	$P_3(s)$	$P_4(s)$	$P_5(s)$	$\bar{y}$	$e=\bar{y}-y$
0	0	1.300		1	9	6	42	18	6	1.310	10
1	0.3	1.245		1	7	2	-14	-22	-14	1.236	-9
2	0.6	1.095		1	5	-1	-35	-17	1	1.098	3
3	0.9	0.855		1	3	-3	-31	3	11	0.868	13
4	1.2	0.514		1	1	-4	-12	18	6	0.514	0
5	1.5	0.037		1	-1	-4	12	18	-6	0.017	-20
6	1.8	-0.600		1	-3	-3	31	3	-11	-0.602	-2
7	2.1	-1.295		1	-5	-1	35	-17	-1	-1.263	32
8	2.4	-1.767		1	-7	2	14	-22	14	-1.793	-26
9	2.7	-1.914		1	-9	6	-42	18	-6	-1.908	6
			(A)	10	330	132	8580	2860	780		
			(B)	-.530	66.802	-7.497	-41.659	14.515	-1.627		
			(C)	-.0530	.20243	-.05680	-.00486	.00508	-.00209		

The values of x and y are arranged in parallel columns and beside them is placed the table of P's for $n = 9$. In the line marked (A) is placed the sum of the squares of the entries in the column above. In the line (B) appears the sum of the products of the entries in the column above by the corresponding values of y. Thus for the column P_2 we have

$$6(1.300) + 2(1.245) - 1(1.095) - 3(0.855) - 4(0.514) - 4(0.037) - 3(-0.600) - 1(1.295) + 2(-1.767) + 6(-1.914) = -7.497.$$

In the line (C) is recorded the value of the entry in (B) divided by the entry in (A). The approximating polynomial is

$$Q(s) = -0.0530\, P_0(s) + 0.20243\, P_1(s) - 0.05680\, P_2(s) -0.00486\, P_3(s) + 0.00508\, P_4(s) - 0.00209\, P_5(s),$$

in which the P's denote the tabulated values, not the true values, of $P_{m,n}(s)$. In the column headed $\bar{y}$ are tr entered the values of $\bar{y} = Q(s)$ for $s = 0, 1, 2, \ldots, 9$. Each value of $\bar{y}$ is the sum of products of the table entries in that row by the corresponding entries in (C). Thus for $s = 3$ we have

$$1(-0.0530) + 3(0.20243) - 3(-0.05680) - 31(-0.00486)$$
$$+ 3(0.00508) + 11(-0.00209) = 0.868$$

The differences in the last column between y and $\bar{y}$ give an idea of the closeness of the approximation.

Example 2. In a certain chemical reaction the concentration of the decomposition product was obtained by experiment in terms of time as follows:

Time (Minutes)	Concentration y
0	0
5	0.000127
10	216
15	286
20	344
25	387
30	415
35	437
40	451
45	460
50	466

It is desired to represent y approximately as a polynomial of degree five, using least squares.

s	Time (Minutes)	Concentra. y	$P_0(s)$	$P_1(s)$	$P_2(s)$	$P_3(s)$	$P_4(s)$	$P_5(s)$	$\bar{y}$	$\bar{y} - y$
0	0	0	1	5	15	30	6	3	0.000001	1
1	5	0.000127	1	4	6	-6	-6	-6	125	-2
2	10	216	1	3	-1	-22	-6	-1	217	1
3	15	286	1	2	-6	-23	-1	4	288	2
4	20	344	1	1	-9	-14	4	4	346	1
5	25	387	1	0	-10	0	6	0	385	-2
6	30	415	1	-1	-9	14	4	-4	416	1
7	35	437	1	-2	-6	23	-1	-4	438	1
8	40	451	1	-3	-1	22	-6	1	451	0
9	45	460	1	-4	6	6	-6	6	459	-1
10	50	466	1	-5	15	-30	6	-3	466	0
		(A)	11	110	858	4290	286	156		
		(B)	3589	-4740	-5194	-2345	-93	-53	$\times 10^{-6}$	
		(C)	326273	-43091	-6054	-547	-325	-340	$\times 10^{-9}$	

Exercises

1. Approximate $\frac{1}{1 + x}$ at $x = 0, 1, \ldots, 8$ by a polynomial of the 4th degree.
2. Approximate sin x at $x = 10^\circ, 15^\circ, 20^\circ, 25^\circ, 30^\circ, 35^\circ, 40^\circ$ by a polynomial of 3rd degree.
3. Approximate the following data by a polynomial of the 5th degree:

x:	0	1	2	3	4	5	6	7	8
y:	1	.9950	.9798	.9539	.9165	.8660	.8000	.7141	.6000.

4. Approximate the following data by a polynomial of the 2nd degree:

x:	0	1	2	3	4	5	6	7	8	9	10.
y:	3	87	156	210	238	252	239	211	158	90	-5

5. Show that

$$P_{m,n}(n - x) = (-1)^m P_{m,n}(x).$$

6. Show that if a fit has been made with an mth degree polynomial and it is decided to use a higher degree only the additional coefficients need be computed, as those already calculated remain unchanged.
7. Show that the fit may be made better, but never worse, by increasing the degree.
8. Show that if $m = n$ the fit is perfect at the given points and the least square polynomial is identical with the interpolating polynomial.

73. GRADUATION, OR SMOOTHING OF DATA

The formulas and tables developed in Article 72 may be applied to the problem of smoothing or "graduating" observed data, when these data are observations of equally spaced values of some unknown function f(x). Let us assume that the interval between the equally spaced values is small enough that f(x) can be represented over several consecutive values with sufficient accuracy by a polynomial of degree m. Then by Least Squares, we fit a polynomial of degree m to $n + 1$ consecutive values obtained by observation, where $n > m$, and use the values calculated from the fitted polynomial in place of the observed values. The table of values, or the curve, thus obtained for f(x) will in general be "smoother" than the original values given by observation.

In practice it is convenient to take the degree m to be odd and the number $n + 1$ of points also odd. Moreover, it is convenient for reasons of symmetry to select the midpoint of the range of $n + 1$ points as the one whose value will be calculated from the approximating polynomial. To illustrate, let us take the degree $m = 3$ and five points, so that $n = 4$. The midpoint is then $x = 2$. Let y_0, y_1, y_2, y_3, y_4 be five consecutive observed values of f(x) and let y_2' be the value obtained for the midpoint from the approximating polynomial. Then putting $m = 3$, $n = 4$, $x = 2$ in equation (10) of of Article 72 we have

$$y_2' = \sum_{k=0}^{3} \left[\frac{\sum_{j=0}^{4} P_{k,4}(j) y_j}{S_k} \right] P_{k,4}(2),$$

and by interchanging the summations with respect to j and k we can rewrite the equation in the form

$$y_2' = \sum_{j=0}^{4} y_j \left[\sum_{k=0}^{3} \frac{P_{k,4}(j) P_{k,4}(2)}{S_k} \right].$$

Now, the quantities in the square brackets are independent of the observed values y_j and may be computed once for all. They are obtained from the tabulated values of $P_{k,n}(x)$ and when computed give

$$(1) \qquad y_2' = \frac{1}{35} \left[-3y_0 + 12y_1 + 17y_2 + 12y_3 - 3y_4 \right].$$

This is an elementary type of smoothing formula.

Example 1. In the tabulation below y represents the number of students who made a grade x on a certain placement examination in mathematics. The values of y' are those obtained by an application of the smoothing formula (1). It is assumed in applying the formula that $y = 0$ for $x = -2, -1,$ and for $x = 21, 22$. The calculation of each y' is done quite simply on a machine by accumulating the algebraic sum of products and then dividing by 35.

The values of y" are those obtained by applying formula (1) to the values of y'. This process may be repeated to obtain still smoother values, but must not be repeated too many times as the essential trend of the original values will eventually be obliterated.

x	y	y'	y"
0	0	-1.4	-0.8
1	13	17.3	19.4
2	69	70.5	74.4
3	147	148.5	143.9
4	208	195.6	196.6
5	195	209.5	203.9
6	195	175.8	180.2
7	126	145.8	143.4
8	130	119.7	126.7
9	118	125.3	118.4
10	121	109.3	112.3
11	85	98.6	97.3
12	93	85.0	87.0
13	75	75.9	73.3
14	54	55.8	56.5
15	42	39.9	41.8
16	30	35.1	33.4
17	34	25.9	26.5
18	10	16.6	15.3
19	8	4.7	6.6
20	1	2.3	1.2

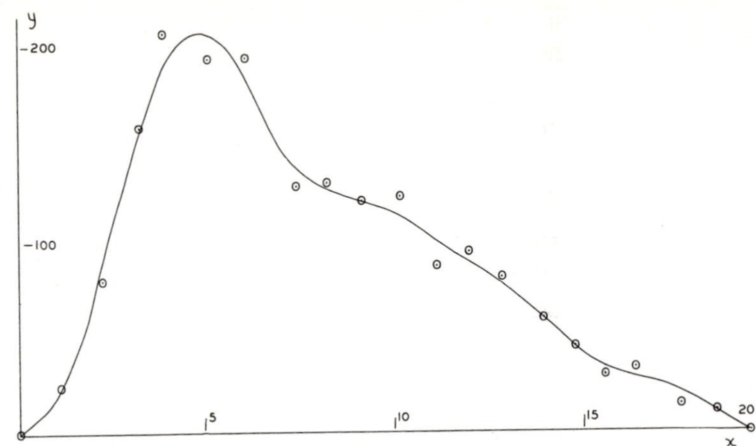

The effect of smoothing is exhibited in the accompanying graph where the original values of y are plotted as circled points and y" is shown as a continuous curve. The average of the absolute values of the fourth differences for y is about 99, for y' about 35, and for y" about 11. This may be interpreted to indicate that y' is a smoother function than y, and that y" is smoother than y', since our use of a third degree polynomial implies that for perfect smoothness, fourth differences should vanish.

The procedure illustrated above for $m = 3$, $n = 4$, may be carried out for any even value of n and any odd value of m, $m < n$. In this way we can derive a whole series of formulas, a few of which follow. (For brevity the y's are omitted and only the coefficients are written.)

1. Smoothing Formulas based on Least Square Approximation by <u>third</u> degree polynomials.

(7 points) $y' = \frac{1}{21}\left[-2 + 3 + 6 + 7 + 6 + 3 - 2\right]$

(9 points) $y' = \frac{1}{231}\left[-21 + 14 + 39 + 54 + 59 + 54 + 39 + 14 - 21\right]$

(11) points) $y' = \frac{1}{429}\left[-36 + 9 + 44 + 69 + 84 + 89 + \text{etc.}\right]$

(In this and the following formulas the omitted coefficients are supplied by symmetry.)

(13 points) $y' = \frac{1}{143}\left[-11 + 0 + 9 + 16 + 21 + 24 + 25 + \text{etc.}\right]$

(15 points) $y' = \frac{1}{1105}\left[-78 - 13 + 42 + 87 + 122 + 147 + 162 + 167 + \text{etc.}\right]$

(17 points) $y' = \frac{1}{323}\left[-21 - 6 + 7 + 18 + 27 + 34 + 39 + 42 + 43 + \text{etc.}\right]$

(19 points) $y' = \frac{1}{2261}\left[-136 - 51 + 24 + 89 + 144 + 189 + 224 + 249 + 264 + 269 + \text{etc.}\right]$

(21 points) $y' = \frac{1}{3059}(-171 - 76 + 9 + 84 + 149 + 204 + 249 + 284 + 309 + 324 + 329 + \text{etc.})$

II. Smoothing Formulas based on Least Square Approximation by <u>fifth</u> degree polynomials.

(7 points) $y' = \frac{1}{231}\left[5 - 30 + 75 + 131 + 75 - 30 + 5\right]$

(9 points) $y' = \frac{1}{429}\left[15 - 55 + 30 + 135 + 179 + 135 + 30 - 55 + 15\right]$

(11 points) $y' = \frac{1}{429}\left[18 - 45 - 10 + 60 + 120 + 143 + \text{etc.}\right]$

(In this and the following formulas the omitted coefficients are supplied by symmetry.)

(13 points) $y' = \frac{1}{2431}\Big[110 - 198 - 135 + 110 + 390 + 600 + 677 + \text{etc.}\Big]$

(15 points) $y' = \frac{1}{46189}\Big[2145 - 2860 - 2937 - 165 + 3755 + 7500 + 10125 + 11063 + \text{etc.}\Big]$

(17 points) $y' = \frac{1}{4199}\Big[195 - 195 - 260 - 117 + 135 + 415 + 660 + 825 + 883 + \text{etc.}\Big]$

(19 points) $y' = \frac{1}{7429}\Big[340 - 255 - 420 - 290 + 18 + 405 + 790 + 1110 + 1320 + 1393 + \text{etc.}\Big]$

(21 points) $y' = \frac{1}{260015}\Big[11628 - 6460 - 13005 - 11220 - 3940 + 6378 + 17655 + 28190 + 36660 + 42120 + 44003 + \text{etc.}\Big]$

74. ALTERNATIVE TREATMENT OF SMOOTHING FORMULAS

The subject of smoothing may be considered from a point of view somewhat different in appearance from that of the foregoing article. Suppose that the "true" values of the desired function are u_i, the observed values are y_i and that

$$y_i = u_i + e_i,$$

where the e_i are random errors. We assume that:

1. The 2m-th differences of the u_i vanish.
2. The number N of ordinates y_i is large compared to m.
3. The errors e_i and e_j are independent if $i \neq j$.
4. For fixed i but for many repeated observations, the values of e_i are distributed according to the normal law of errors with small standard deviation, σ.

With these hypotheses, it may be assumed that sums

of the type $\sum e_i e_j$ are small and may be neglected in comparison with sums of the type $\sum e_i^2$.

It will be convenient also to assume that when values of the e_i beyond the range of given values occur in our formulas they may be replaced by zero. With this understanding sums of the form $\sum e_i^2$, $\sum e_{i+1}^2$, etc., will all be equal.

Let

$$(1) \quad y_i' = y_i + a_0 \, \delta^{2m} y_i + a_1 \, \delta^{2m+2} y_i + \cdots + a_k \, \delta^{2m+2k} y_i.$$

Note that the differences involved are even-ordered central differences and when expressed in terms of ordinates are symmetrical with respect to the central ordinate y_i. Also let

$$y' = u_i + e_i'$$

so that e_i' is the error of y_i'. In equation (1) we replace y_i by $u_i + e_i$, y_i' by $u_i + e_i'$, and find that

$$(2) \quad e_i' = e_i + a_0 \, \delta^{2m} e_i + a_i \, \delta^{2m+2} e_i + \cdots + a_k \, \delta^{2m+2k} e_i,$$

since by assumption the 2m-th and all higher differences of u_i vanish.

Now the a's are to be chosen so as to minimize

$$E' = \sum (e_i')^2.$$

Proceeding in the usual manner to find the a's we get k + 1 equations

$$(3) \quad \sum \delta^{2m+2j} e_i \left[e_i + a_o \delta^{2m} e_i + \ldots + a_k \delta^{2m+2k} e_i \right] = 0 \quad (j = 0, 1, \ldots, k)$$

These can be simplified if we note a) that sums of all cross products of the e's are negligible because of 1 - 4; b) that all sums of squares of the form $\sum e_{i+j}^2$ are equal to $\sum e_i^2$ which we denote by E; c) that in view of a) and b) $\sum \left(\delta^{2m+2p} e_i\right) \left(\delta^{2m+2q} e_i\right) = CE$ in which C is the sum of products of the corresponding binomial coefficients with alternating signs of degrees (2m + 2p) and (2m + 2q) lined up so that the central coefficients correspond. If the degrees are 4 and 6, for example, the sum of products is

$$\begin{array}{l} 1 - 6 + 15 - 20 + 15 - 6 + 1 \\ 0 + 1 - 4 + 6 - 4 + 1 + 0 \\ \hline 0 - 6 - 60 - 120 - 60 - 6 + 0 = -252 = C \end{array}$$

In general the sum of such products is

$$C = (-1)^{p-q} \binom{4m+2p+2q}{2m+p+q} .$$

With these facts in mind we may write the equations (3) as

$$(4) \quad (-1)^{m+j} \binom{2m+2j}{m+j} + a_o (-1)^j \binom{4m+2j}{2m+j} + a_1 (-1)^{j+1} \binom{4m+2j+2}{2m+j+1} + \ldots + a_k (-1)^{j+k} \binom{4m+2j+2k}{2m+j+k} = 0, \quad (j = 0, 1, \ldots, k).$$

In particular if $m = 2$, $k = 1$, the equations (4) are

$$\binom{4}{2} + a_0\binom{8}{4} - a_1\binom{10}{5} = 0,$$

$$-\binom{6}{3} - a_0\binom{10}{5} + a_1\binom{12}{6} = 0,$$

which, with the values of the binomial coefficients substituted become

$$6 + 70a_0 - 252a_1 = 0,$$

$$-20 - 252a_0 + 924a_1 = 0,$$

giving

$$a_0 = -\frac{3}{7} \qquad a_1 = \frac{-2}{21}.$$

The desired smoothing formula is therefore

$$y_i' = y_i - \frac{3}{7}\delta^4 y_i - \frac{2}{21}\delta^6 y_i \tag{5}$$

for third degree approximation using 7 points. If we substitute in (5) the values of the differences in terms of ordinates, we find that (5) is identical with the third-degree formula for seven points given in Article 73. In fact all of the formulas of Article 73 may be derived by the present method. In this manner we obtain the additional formulas for seventh degree approximation

(6) (9 points) $y_i' = y_i - \frac{7}{1287}\delta^8 y_i$

(7) (11 points) $y_i' = y_i - \frac{7}{143}\delta^8 y_i - \frac{28}{2431}\delta^{10} y_i.$

When the values of the a's obtained from (4) are put

back in (2) and the value of $E' = \sum(e_i')^2$ is computed, it will be found by use of (4) that the result reduces to

$$E' = E\left[1 + (-1)^m \binom{2m}{m} a_0 + (-1)^{m+1} \binom{2m+2}{m+1} a_1 + \cdots + (-1)^{m+k} \binom{2m+2k}{m+k} a_k\right].$$

Examination of the expression in brackets will reveal that it is exactly the midcoefficient of the smoothing formula expressed in terms of ordinates.

Hence the result:

$$E' = E(\text{midcoefficient}).$$

In words, this states that "The sum of squares of the errors of the smoothed values equals the sum of squares of the errors of the original values times the midcoefficient of the smoothing formula."

For convenience of reference we list the approximate values of the midcoefficients for the formulas of Article 73, and for formulas (6) and (7) of the present article.

No. Points	5	7	9	11	13	15	17	19	21
3rd degree	0.49	0.33	0.26	0.21	0.17	0.15	0.13	0.12	0.11
5th degree	—	0.57	0.42	0.33	0.28	0.24	0.21	0.19	0.17
7th degree	—	—	0.62	0.48	—	—	—	—	—

It is apparent that the overall accuracy of the smoothed values increases as the number of points increases and decreases as the degree increases. Hence, in choosing a smoothing formula the degree should be taken as low as possible consistent with the basic hypothesis that the 2m-th differences of the true values are negligible, and the number of points as large as is practical, in view of the total range of given values, increased labor of computation, and increased accumulation of "rounding off" errors. In the final analysis much depends on the good judgment of the computer.

Exercises

1. Express formulas (6) and (7) in terms of ordinates.
2. Derive a smoothing formula of 9th degree for 11 points.
3. Apply the 5th degree, eleven-point formula to the data of the example in Article 73, and compare results.

75. GAUSS'S METHOD OF NUMERICAL INTEGRATION

An important method of numerical integration due to Gauss has been deferred to this point because its derivation involves the use of the Legendre polynomials obtained in Article 70.

In Chapter IV various formulas for numerical integration were derived on the assumption that the points $x_0, x_1, \ldots, x_n$ were given and that the coefficients were to be determined so as to make the formula valid for all polynomials of as high degree as possible. In Gauss's formulas both the x's and the coefficients are treated as unknowns to be determined so as to make the formula valid for all polynomials of as high degree as possible. It is convenient first of all to change the variable so that the interval of integration with respect to x is from 0 to 1.

Accordingly, let us try to determine not only the coefficients $A_1, A_2, \ldots, A_n$ but also the values of the x's, $x_1, x_2, \ldots, x_n$, in such a way that the formula

$$(1) \quad \int_0^1 y\,dx = A_1\,y(x_1) + A_2\,y(x_2) + \cdots + A_n\,y(x_n)$$

shall be exact when y is any polynomial of as high degree

as possible. A count of the constants available, n A's, and n x's, totalling 2n, suggests that the highest degree will probably be 2n - 1. Hence we let $y = 1, x, x^2, \ldots, x^{2n-1}$ in turn in (1), obtain 2n equations

$$(2) \qquad \int_0^1 x^k \, dx = A_1 x_1^k + A_2 x_2^k + \cdots + A_n x_n^k,$$

$$(k = 0, 1, \ldots, 2n - 1)$$

and endeavor to solve these equations for the n x's and n A's. As these equations are not linear in the x's their solution presents a nasty problem in algebra, which however is neatly handled by the following trick. Multiply the kth equation in (2) by $(-1)^k \binom{n}{k} \binom{n+k}{k}$, $(k = 0, 1, \ldots, n)$ and add. The result is

$$\int_0^1 P_n(x) \, dx = A_1 P_n(x_1) + A_2 P_n(x_2) + \cdots + A_n P_n(x_n).$$

in which $P_n(x)$ is the Legendre polynomial of degree n derived in Article 70. In similar fashion, starting with the second, instead of the first of equations (2), and adding up to n + 1 we get

$$\int_0^1 x P_n(x) \, dx = A_1 x_1 P_n(x_1) + A_2 x_2 P_n(x_2) + \cdots + A_n x_n P_n(x_n)$$

and in general

$$(3) \qquad \int_0^1 x^q P_n(x) \, dx = A_1 x_1^q P_n(x_1) + A_2 x_2^q P_n(x_2) + \cdots + A_n x_n^q P_n(x_n)$$

$q = 0, 1, \ldots, n - 1$. Now the left hand member of (3) vanishes for $q = 0$ to $n - 1$ by the orthogonal property of $P_n(x)$. The right hand members will also vanish, no matter what A's are chosen if $x_1, x_2, \dots, x_n$ are the n roots of $P_n(x) = 0$. These are known to be all real, distinct, and all in the interval from 0 to 1. Accordingly, we choose the x's as the n roots of $P_n(x) = 0$, substitute these values in the first n equations (2), and solve these equations, which are linear in the A's, for the n coefficients A. That these equations will always have a unique solution follows from the fact that the x's are all distinct.

The roots of $P_n(x)$ are for the most part irrational, a fact which detracts somewhat from the practical convenience and usefulness of Gauss's formulas. These formulas however yield higher accuracy in proportion to the number of points utilized than do other quadrature formulas, and this is sometimes an important consideration.

For convenience of reference the roots of $P_n(x) = 0$ and the corresponding coefficients are listed below from $n = 2$ to $n = 9$. The corresponding formulas for integration, obtained by putting these values in (2), are exact for all polynomials of the degree indicated.

Gauss's Numbers

i	x_i	A_i	i	x_i	A_i
	3rd degree			13th degree	
1	.2113249	.5000000	1	.0254460	.0647425
2	.7886751	.5000000	2	.1292344	.1398527
			3	.2970774	.1909150
	5th degree		4	.5000000	.2089796
1	.1127017	.2777778	5	.7029226	.1909150
2	.5000000	.4444444	6	.8707656	.1398527
3	.8872983	.2777778	7	.9745540	.0647425
	7th degree			15th degree	
1	.0694318	.1739274	1	.0198551	.0506143
2	.3300095	.3260726	2	.1016668	.1111905
3	.6699905	.3260726	3	.2372338	.1568533
4	.9305682	.1739274	4	.4082827	.1813418
			5	.5917173	.1813418
	9th degree		6	.7627662	.1568533
1	.0469101	.1184634	7	.8983332	.1111905
2	.2307653	.2393144	8	.9801449	.0506143
3	.5000000	.2844444			
4	.7692347	.2393144		17th degree	
5	.9530899	.1184634	1	.0159199	.0406372
			2	.0819844	.0903241
	11th degree		3	.1933143	.1303053
1	.0337652	.0856622	4	.3378733	.1561735
2	.1693953	.1803808	5	.5000000	.1651197
3	.3806904	.2339570	6	.6621267	.1561735
4	.6193096	.2329570	7	.8066857	.1303053
5	.8306047	.1803808	8	.9180156	.0903241
6	.9662348	.0856622	9	.9840801	.0406372

Example. Find the integral of $\sqrt{1 + 2x}$ from 0 to 1 and check.

(a) Using the 3rd degree formula we have

i	$\sqrt{1 + 2x_i}$	A_i
1	1.1927488	.5
2	1.6054128	.5

Answer 1.3990808

Correct value 1.3987175

Error .0003633

(b) Using the 5th degree formula we have

1	1.1069794	.2777778
2	1.4142136	.4444444
3	1.6657120	.2777778

Answer 1.3987314

Error .0000139

(c) Using the 7th degree formula we have

1	1.0671755	.1739274
2	1.2884172	.3260726
3	1.5296996	.3260726
4	1.6914894	.1739274

Answer 1.3987181

Error .0000006

Exercises

1. $\int_0^1 \sqrt{1 + x^2}\, dx$ (5th degree)

2. $\int_1^2 \sqrt{x}\, dx$ (5th degree)

3. $\int_1^2 \frac{dx}{x}$ (7th degree)

4. Compare error in 1, 2, 3, above with error of Simpson's rule using four intervals. Using ten intervals.

5. Outline a procedure for calculating log x from the integral

$$\log x = \int_1^x \frac{dx}{x}$$

Chapter X

OTHER APPROXIMATIONS BY LEAST SQUARES*

In Chapter IX various forms of polynomial approximations were based on the idea of least squares. We now consider the more general case where the approximating function is not necessarily a polynomial. Again, there are two cases, the continuous case where the integral of the square of the error is minimized, and the discrete case where the sum of the squares of the errors is minimized.

76. GENERAL PROBLEM OF APPROXIMATION BY LEAST SQUARES

In this article and the next we take only the continuous case, since the treatment of the discrete case is identical except for sums instead of integrals.

Let $v_1(x)$, $v_2(x)$, . . ., $v_m(x)$ denote a set of linearly independent continuous functions of x in an interval $a \leqq x \leqq b$. Let w(x) be a continuous non-negative function also defined in (a, b); w(x) is called the "weight" function. Let z(x) be a continuous function in (a, b) which we desire to represent approximatdy by a sum of the form

$$y(x) = \sum_{i=1}^{m} x_i \, v_i(x). \tag{1}$$

The a's in (1) are to be determined in such a manner as to minimize the integral

$$I = \int_a^b w(z - \sum_{i=1}^{m} a_i \, v_i)^2 \, dx. \tag{2}$$

*See Appendix C.

When we differentiate I with respect to $a_1, a_2, \ldots, a_m$, and equate all the partial derivatives to zero, we obtain m equations for the determination of the a's:

$$\int_a^b w\, v_k(z - \sum a_i\, v_i)dx = 0, \qquad (3)$$

$$k = 1, 2, \ldots, m.$$

It may be shown that under the assumptions we have made, the equations (3) possess a unique solution for the a's and that the values so obtained actually minimize the integral I.

We have already noted for the polynomial case the simplification which occurs when the approximating set of functions is orthogonal. In this more general case the set of functions v_i is said to be orthogonal with respect to the weight function w if

$$\int_a^b w\, v_k\, v_i\, dx = 0 \quad \text{if } i \neq k. \qquad (4)$$

When the relation (4) holds, equation (3) is readily solved for the a's and we find

$$a_k = \frac{\int_a^b w\, z\, v_k\, dx}{\int_a^b w\, {v_k}^2\, dx} \qquad (5)$$

77. FUNDAMENTAL THEOREM

One of the most frequent uses of approximation by Least Squares occurs in the case where the given function z(x) has been obtained from observations or experiments subject to random errors. There is assumed to exist a "true" function u(x) which would have been found if it had been possible to make the observations complete or to perform the experiments without error. It is this true function u(x) which we wish to discover, not just a more or less adequate expression for the observed function z(x). The following theorem gives us some useful information about the true function.

Theorem. *If the true function* u(x) *can be represented by the* v's, *so that*

$$u = b_1 v_1 + b_2 v_2 + \cdots + b_m v_m,$$

and if the a's *are determined so that*

$$y = a_1 v_1 + a_2 v_2 + \cdots + a_m v_m$$

is the Least Square approximation to the observed function z(x), *then*

$$\int_a^b w(z - u)^2\, dx = \int_a^b w(z - y)^2\, dx + \int_a^b w(y - u)^2\, dx$$

Proof. We have

$$\begin{aligned}(z - u)^2 &= \left[(z - y) + (y - u)\right]^2 \\ &= (z - y)^2 + 2(z - y)(y - u) \\ &\qquad + (y - u)^2\end{aligned}$$

$$z - y = z - \sum_{i=1}^{m} a_i v_i$$

$$y - u = \sum_{k=1}^{m} (a_k - b_k) v_k$$

whence

$$(z - y)(y - u) = \sum_{k=1}^{m} (a_k - b_k) v_k \left(z - \sum_{i=1}^{m} a_i v_i\right).$$

Hence

$$(z - u)^2 = (z - y)^2 + (y - u)^2 + 2 \sum_{k=1}^{m} (a_k - b_k) v_k \, (z - \sum_{i=1}^{m} a_i v_i).$$

Multiplying this equation by w dx and integrating from a to b we get

$$\int_a^b w(z - u)^2 \, dx = \int_a^b w(z - y)^2 \, dx + \int_a^b w(y - u)^2 \, dx$$

since all of the integrals

$$\int_a^b w \, v_k (z - \sum_{i=1}^{m} a_i v_i) \, dx$$

vanish because of equation (3) of Article 76.

The theorem tells us that (excepting the trivial case where $y \equiv z$)

$$\int_a^b w(y - u)^2 \, dx < \int_a^b w(z - u)^2 \, dx.$$

This inequality may be interpreted to say, "The function y obtained as an approximation to the observed function z is closer to the true function u than is the observed function z itself."

It must not be overlooked that in making this assertion we are assuming that the true function can be represented by the v's. As stated in the previous article, this result applies to the case of sums just as well as to integrals.

78. TRIGONOMETRIC APPROXIMATION

Historically one of the earliest, and in its influence on the course of mathematical research the most significant of all, is the theory of approximation by means of trigonometric sums and the related Fourier Series. Such approximation is particularly suited to the representation of periodic functions, though it occurs also in many situations where periodicity is not obviously essential.

Let f(x) denote a continuous function of x which is periodic with period p, so that

$$f(x + p) = f(x).$$

It is convenient to change the variable x so that the period will be 2π. This may be accomplished by the simple substitution $x = (p/2\pi)x'$ where x' is the new variable. We shall henceforth assume that this transformation has already been effected and that the period of f(x) is 2π.

We desire to determine the coefficients a_m, b_m in the trigonometric sum

$$\sum_{m=0}^{n}(a_m \cos mx + b_m \sin mx)$$

in such a way as to minimize the integral

$$(1) \qquad E = \int_{-\pi}^{\pi} \left[f(x) - \sum(a_m \cos mx + b_m \sin mx)\right]^2 dx$$

The procedure is the same as that used in the case of polynomial approximations by Least Squares. We differentiate E with respect to a_k and b_k, $(k = 0, 1, \ldots, n)$, equate the partial derivatives to zero and after slight simplification arrive at the equations

$$(2) \qquad \sum_{m=0}^{n}(a_m \int_{-\pi}^{\pi} \cos mx \ \cos kx\, dx +$$

$$b_m \int_{-\pi}^{\pi} \sin mx \cos kx\, dx) = \int_{-\pi}^{\pi} f(x) \cos kx\, dx$$

$$(3)\qquad \sum_{m=0}^{n}\left(a_m \int_{-\pi}^{\pi} \cos mx \sin kx\, dx + b_m \int_{-\pi}^{\pi} \sin mx \sin kx\, dx\right) = \int_{-\pi}^{\pi} f(x) \sin kx\, dx$$

$$(k = 0, 1, \ldots, n).$$

It is readily verified by integration that when m and k are integers

$$(4)\qquad \int_{-\pi}^{\pi} \cos mx \sin kx\, dx = \int_{-\pi}^{\pi} \sin mx \cos kx\, dx = 0$$

$$(5)\qquad \int_{-\pi}^{\pi} \cos mx \cos kx\, dx = \begin{cases} 0 & \text{if } k \neq m \\ \pi & \text{if } k = m \neq 0 \\ 2\pi & \text{if } k = n = 0 \end{cases}$$

$$(6)\qquad \int_{-\pi}^{\pi} \sin mx \sin kx\, dx = \begin{cases} 0 & \text{if } k \neq m \\ \pi & \text{if } k = m \neq 0. \end{cases}$$

These equations show that the functions 1, cos x, sin x, cos 2x, sin 2x, cos 3x, sin 3x, . . ., form an orthogonal set for the interval $-\pi < x < \pi$. Substituting the values of the integrals from (4), (5), and (6) into (2) and (3) and solving for a_k and b_k we obtain

$$(7)\qquad a_0 = \frac{1}{2\pi} \int_{-\pi}^{\pi} f(x)\, dx,$$

$$(8)\qquad a_k = \frac{1}{\pi} \int_{-\pi}^{\pi} f(x) \cos kx\, dx \quad (k = 1, 2, \ldots,),$$

$$(9)\qquad b_k = \frac{1}{\pi} \int_{-\pi}^{\pi} f(x) \sin kx\, dx \quad (k = 1, 2, \ldots,).$$

These equations give the values of the coefficients which

we set out to determine. They are called the "Fourier" coefficients.

Two special cases deserve mention. a) If $f(x)$ is an <u>even</u> function,

$$f(x) = f(-x),$$

we see from (7), (8), and (9) that

$$a_0 = \frac{1}{\pi} \int_0^{\pi} f(x)\, dx,$$

$$a_k = \frac{2}{\pi} \int_0^{\pi} f(x) \cos kx\, dx,$$

$$b_k = 0$$

so that all the sines drop out and $f(x)$ will be represented by cosine terms only. b) If $f(x)$ is odd, $f(x) = -f(-x)$, and

$$a_0 = 0, \qquad a_k = 0$$

$$b_k = \frac{2}{\pi} \int_0^{\pi} f(x) \sin kx\, dx.$$

In this case $f(x)$ is represented by sines only.

<u>Example 1</u>. Obtain the trigonometric approximation to $f(x) = x(\pi^2 - x^2)$ in the interval $-\pi$ to π.

Since the function $x(\pi^2 - x^2)$ as it stands is obviously not periodic, some explanation is required. The key to the difficulty lies in the fact that $f(x) = x(\pi^2 - x^2)$ <u>only</u> in the interval $-\pi \leq x \leq \pi$. In intervals of length 2π <u>outside</u> this interval $f(x)$ is to be defined by the periodic property $f(x + 2\pi) = f(x)$.

For example, in the interval $\pi \leqq x \leqq 3\pi$ we would have $f(x) = (x - 2\pi)\left[\pi^2 - (x - 2\pi)^2\right]$.

With this interpretation we see that f(x) is periodic of period 2π, and also is an odd function, so that its trigonometric approximation will contain sines only. Using the formula derived for b_k in this case we have

$$b_k = \frac{2}{\pi} \int_0^{\pi} x(\pi^2 - x^2) \sin kx \, dx.$$

One integration by parts gives

$$b_k = \frac{2}{k\pi} \int_0^{\pi} (\pi^2 - 3x^2) \cos kx \, dx$$

since the integrated portion vanishes at both limits. A second integration by parts yields

$$b_k = \frac{12}{k^2\pi} \int_0^{\pi} x \sin kx \, dx$$

where again the integrated portion vanishes because of the factor sin kx. Further integration gives finally

$$b_k = -\frac{12}{k^3} \cos k\pi = -\frac{(-1)^k 12}{k^3}.$$

The desired trigonometric representation of f(x) is therefore

$$y = 12\left[\sin x - \frac{\sin 2x}{8} + \frac{\sin 3x}{27} - \frac{\sin 4x}{64} + \cdots - \frac{(-1)^n \sin nx}{n^3}\right].$$

<u>Example 2</u>. Obtain the trigonometric approximation

to f(x) where

$$f(x) = \begin{cases} x^2(3\pi - 2x) & \text{if} \quad 0 \leq x < \pi \\ x^2(3\pi + 2x) & \text{if} \quad -\pi < x \leq 0. \end{cases}$$

The function f(x) is again defined outside of the given interval by the periodic property. Since f(x) is even

$$a_0 = \frac{1}{\pi} \int_0^{\pi} f(x)\, dx,$$

$$a_k = \frac{2}{\pi} \int_0^{\pi} f(x) \cos kx\, dx.$$

These integrals are evaluated as before and we find

$$a_0 = \pi^3/2$$

$$a_k = \frac{-48}{\pi k^4} \qquad \text{if } k \text{ is odd,}$$

$$a_k = 0 \qquad \text{if } k \text{ is even.}$$

Consequently the desired representation is

$$y = \frac{\pi^3}{2} - \frac{48}{\pi}\left[\cos x + \frac{\cos 3x}{3^4} + \frac{\cos 5x}{5^4} + \frac{\cos 7x}{7^4} + \cdots\right]$$

An adequate discussion of the error committed in terminating the trigonometric sum with terms of order n (i.e., terms involving cos nx, sin nx) is beyond the scope of an elementary treatment.

Exercises

Assuming f(x) periodic of period 2π find the trigonometric approximation.

1. $f(x) = x(\pi - x)$ if $0 \leq x < \pi$
   $= x(\pi + x)$ if $-\pi < x \leq 0.$

2. $f(x) = \pi - 2x$ if $0 \leq x < \pi$

$= \pi + 2x$ if $-\pi < x \leq 0$

3. $f(x) = x^4 - 3\pi x^3 - 2\pi^2 x^2 + 3\pi^3 x + \pi^4 \quad -\pi < x < \pi$.

4. $f(x) = 1 \quad 0 < x < \pi$,

$= -1 \quad -\pi < x < 0$.

5. In Exercise 1, plot f(x) and the first five terms of the approximation on the same axes. Compare the curves.

6. Show by consideration of the second order partial derivatives that the Fourier coefficients actually make E a minimum.

79. HARMONIC ANALYSIS

Instead of determining the coefficients so as to minimize the integral of the square of the difference, we may determine them so as to minimize the sum of the squares of the differences taken at points $x_0, x_1, \ldots, x_p$ in the interval $-\pi < x < \pi$.

Let f(x) denote a periodic function of x of period 2π. Let the interval $-\pi$ to π be divided into 2k equal parts and let the points of division be designated by x_j where

$$x_j = \frac{j\pi}{k}.$$

We propose to minimize the expression

$$E = \sum_{j=-k}^{k-1} \left[f(x_j) - \sum_{m=0}^{n} (a_m \cos mx_j + b_m \sin mx_j) \right]^2$$

Differentiating partially with respect to the a's and b's and setting the partial derivatives equal to zero, we are led to the set of equations

$$(1) \quad \sum_{m=0}^{n} \left[a_m \sum_{j=-k}^{k-1} \cos mx_j \cos qx_j + b_m \sum_{j=-k}^{k-1} \sin mx_j \cos qx_j \right] = \sum_{j=-k}^{k-1} f(x_j) \cos qx_j,$$

$$(2)\quad \sum_{m=0}^{n}\left[a_m \sum_{j=-k}^{k-1} \cos mx_j \sin qx_j + b_m \sum_{j=-k}^{k-1} \sin mx_j \sin qx_j\right] = \sum_{j=-k}^{k-1} f(x_j) \sin qx_j$$

$$(q = 0, 1, 2, \ldots, n).$$

It may be shown by means of suitable trigonometric identities that if m and q are integers not less than zero and not greater than k then

$$\sum_{j=-k}^{k-1} \sin mx_j \cos qx_j = 0$$

$$(3)\quad \sum_{j=-k}^{k-1} \cos mx_j \cos qx_j = \begin{cases} 0 & \text{if } m \neq q \\ k & \text{if } m = q \neq 0 \text{ and } \neq k \\ 2k & \text{if } m = q = 0 \text{ or } = k \end{cases}$$

$$\sum_{j=-k}^{k-1} \sin mx_j \sin qx_j = \begin{cases} 0 & \text{if } m \neq q \\ k & \text{if } m = q \neq 0 \text{ and } \neq k \\ 0 & \text{if } m = q = 0 \text{ or } = k \end{cases}$$

When these values for the sums are put back in equations (1) and (2) we can easily solve for the coefficients and find that

$$(4)\quad a_0 = \frac{1}{2k} \sum_{j=-k}^{k-1} f(x_j),$$

$$(5)\quad a_q = \frac{1}{k} \sum_{j=-k}^{k-1} f(x_j) \cos qx_j, \qquad 0 < q < k.$$

$$(6)\quad b_q = \frac{1}{k} \sum_{j=-k}^{k-1} f(x_j) \sin qx_j.$$

The summations on the right may be put in a more symmetrical form if we observe that due to the periodicity the term for $j = -k$ is equal to the term for $j = k$ and is consequently equal to half the sum of both. Thus the first equation may be written

$$a_0 = \frac{1}{2\pi}\cdot\frac{\pi}{k}\left[\frac{1}{2}f(x_{-k}) + f(x_{-k+1}) + \ldots + f(x_{k-1}) + \frac{1}{2}f(x_k)\right]$$

and similar formulas hold for a_q and b_q. Since $\pi/k = h$ is the length of the interval between equally spaced points, it is now clear that the values of a_0, a_q, b_q obtained in (4), (5), and (6) above are just what we would have secured if we had expressed the integrals in (7), (8), and (9) of Article 78 by means of the Trapezoidal Rule. In this connection it is interesting to observe that the Trapezoidal Rule enjoys a peculiar distinction when applied to a period of a periodic function. In formula (2) p. 198 suppose that the interval from x_0 to x_n is a period of the function $y = f(x)$. Then evidently $\Delta y_{n-1} = \Delta y_{-1}$, $\Delta y_n = \Delta y_0$, $\Delta^3 y_{n-2} = \Delta^3 y_{-2}$, etc., so that the corrective terms all drop out of the formula. We may conclude that the Trapezoidal Rule is more accurate when applied to a period of a periodic function than when applied to a non-periodic function, other things being equal.

80. CALCULATION OF COEFFICIENTS

Just as in the case of integrals it turns out that if $f(x)$ is an even function the b's are all zero and the a's are found by summing over the half interval and doubling the result. The same holds, *mutatis mutandis*, for an odd function. We make use of these facts to arrange the computation so as to reduce the labor of calculating the coefficients by about one half for functions that are neither odd nor even.

Since

$$f(x) = \frac{1}{2}\left[f(x) + f(-x)\right] + \frac{1}{2}\left[f(x) - f(-x)\right]$$

and the first term on the right is even and the second odd we see that

$$a_q = \frac{1}{k}\sum_{j=-k}^{k-1} f(x_j) \cos qx_j$$

$$= \frac{1}{2k}\sum_{j=-k}^{k-1}\Big[f(x_j) + f(-x_j)\Big]\cos qx_j = \frac{1}{k}\sum_{j=0}^{k} F_j \cos qx_j$$

where $F_0 = f(0)$, $F_j = f(x_j) + f(-x_j)$, $F_k = f(x_k)$. Similarly

$$b_q = \frac{1}{k}\sum_{j=1}^{k-1} G_j \sin qx_j,$$

where $G_j = f(x_j) - f(-x_j)$. Hence it saves labor to tabulate F_j and G_j first and then to obtain the a's and b's from these.

Another labor-saving observation is this: if k is even, all the $2(k + 1)^2$ quantities $\cos qx_j$, $\sin qx_j$, $q = 0, 1, 2, \ldots, k$, $j = 0, 1, 2, \ldots, k$, are merely repetitions, with sign changes, of the fundamental set of values $\cos x_j$, $j = 0, 1, \ldots, \frac{n}{2}$.

To illustrate the practical application of these remarks we set up a computational form for the case $k = 8$, where the function $f(x)$ is neither odd nor even. For brevity in writing let $\cos(\pi/8) = a$, $\cos(2\pi/8) = b$, $\cos(3\pi/8) = c$. Then the forms for calculating the coefficients up to the sixth order appear as shown below.

x	F	cos x	cos 2x	cos 3x	cos 4x	cos 5x	cos 6x	
0	$f_0 = F_0$	1	1	1	1	1	1	
$\pi/8$	$f_1 + f_{-1} = F_1$	a	b	c	0	-c	-b	
$2\pi/8$	$f_2 + f_{-2} = F_2$	b	0	-b	-1	-b	0	
$3\pi/8$	$f_3 + f_{-3} = F_3$	c	-b	-a	0	a	b	
$4\pi/8$	$f_4 + f_{-4} = F_4$	0	-1	0	1	0	-1	Form A
$5\pi/8$	$f_5 + f_{-5} = F_5$	-c	-b	a	0	-a	b	
$6\pi/8$	$f_6 + f_{-6} = F_6$	-b	0	b	-1	b	0	
$7\pi/8$	$f_7 + f_{-7} = F_7$	-a	b	-c	0	c	-b	
	$f_8 = F_8$	-1	1	-1	1	-1	1	
	a_0	a_1	a_2	a_3	a_4	a_5	a_6	

x	G	sin x	sin 2x	sin 3x	sin 4x	sin 5x	sin 6x	
$\pi/8$	$f_1 - f_{-1} = G_1$	c	b	a	1	a	b	
$2\pi/8$	$f_2 - f_{-2} = G_2$	b	1	b	0	-b	-1	
$3\pi/8$	$f_3 - f_{-3} = G_3$	a	b	-c	-1	-c	b	
$4\pi/8$	$f_4 - f_{-4} = G_4$	1	0	-1	0	1	0	Form B
$5\pi/8$	$f_5 - f_{-5} = G_5$	a	-b	-c	1	-c	-b	
$6\pi/8$	$f_6 - f_{-6} = G_6$	b	-1	b	0	-b	1	
$7\pi/8$	$f_7 - f_{-7} = G_7$	c	-b	a	-1	a	-b	
		b_1	b_2	b_3	b_4	b_5	b_6	

The coefficient a_0 is given by

$$a_0 = \frac{1}{16} \sum_{i=0}^{8} F_i .$$

For the remaining a's and the b's the number of multiplications involved in finding the sums of products may be reduced by appropriate grouping of terms. For instance it is clear that

$$a_2 = \frac{1}{8}\left[(F_0 - F_4 + F_8) + b(F_1 - F_3 - F_5 + F_7)\right]$$

and that

$$b_6 = \left[(G_6 - G_2) + b(G_1 - G_3 - G_5 + G_7)\right] .$$

It turns out that the cases $k = 6$, 12, 24, etc., permit rather effective grouping of terms, and also include the simple values $\sin 30^\circ = \cos 60^\circ = \frac{1}{2}$. Hence these values of k are generally preferred in setting up problems in harmonic analysis.

For beginners or for occasional computations it is suggested that the generalized forms A and B be employed for the calculation of the coefficients. On the other hand, if a considerable number of harmonic analyses have to be performed, it will pay the computer to use a more specialized computation form which takes full advantage of the grouping of terms referred to above. Detailed forms for this purpose are given, e.g., by Whittaker and Robinson, The Calculus of Observations, Blackie and Son, Ltd., London and Glasgow, Chap. X, or by Running, Empirical Formulas, John Wiley and Sons, New York, Chap. V.

If the function f(x) is even, the function G is zero and $F_0 = f_0$, $F_1 = 2f_1$, $F_2 = 2f_2$, . . ., $F_k = f_k$. In this case Form A gives the desired coefficients. If f(x) is odd the F is zero and $G_i = 2f_i$, so that Form B is used.

Example 1. If f(x) has the values given and is odd, obtain the trigonometric approximation:

x :	0	30°	60°	90°	120°	150°	180°
f(x):	0	5	8	9	8	5	0

The solution, following the model of Form B appears below. Since the function as given is already odd it is unnecessary to introduce the function G.

x	f	sin x	sin 2x	sin 3x	sin 4x	sin 5x
0	0	0	0	0	0	0
30°	5	$\frac{1}{2}$	$\frac{1}{2}\sqrt{3}$	1	$+\frac{1}{2}\sqrt{3}$	$\frac{1}{2}$
60°	8	$\frac{1}{2}\sqrt{3}$	$\frac{1}{2}\sqrt{3}$	0	$-\frac{1}{2}\sqrt{3}$	$-\frac{1}{2}\sqrt{3}$
90°	9	1	0	-1	0	+1
120°	8	$\frac{1}{2}\sqrt{3}$	$-\frac{1}{2}\sqrt{3}$	0	$\frac{1}{2}\sqrt{3}$	$-\frac{1}{2}\sqrt{3}$
150°	5	$\frac{1}{2}$	$-\frac{1}{2}\sqrt{3}$	1	$-\frac{1}{2}\sqrt{3}$	$\frac{1}{2}$
180°	0	0	0	0	0	0
		$14 + 8\sqrt{3}$	0	1	0	$14 - 8\sqrt{3}$

$$y = \frac{1}{3}\left[(14 + 8\sqrt{3}) \sin x + \sin 3x + (14 - 8\sqrt{3}) \sin 5x\right]$$

$$y = 9.285 \sin x + 0.333 \sin 3x + 0.048 \sin 5x.$$

Example 2. Obtain the trigonometric representation to terms of order six for f(x), where f(x) has the values

j	$f(x_j)$	j	$f(x_j)$	j	$f(x_j)$
-12	4.00	-4	8.42	5	8.10
-11	4.08	-3	9.07	6	7.59
-10	4.22	-2	9.35	7	7.00
-9	4.45	-1	9.53	8	6.34
-8	4.85	0	9.55	9	5.56
-7	5.45	1	9.46	10	4.80
-6	6.40	2	9.25	11	4.21
-5	7.45	3	8.96	12	4.00
		4	8.58		

We set up the functions F_i and G_i and carry out the calculation as indicated below:

$a = 0.96593,\quad b = 0.86603,\quad c = 0.70711,\quad d = 0.25882.$

j	F_j	cos x	cos 2x	cos 3x	cos 4x	cos 5x	cos 6x
0	9.55	1	1	1	1	1	1
1	18.99	a	b	c	$\frac{1}{2}$	d	0
2	18.60	b	$\frac{1}{2}$	0	$-\frac{1}{2}$	-b	-1
3	18.04	c	0	-c	-1	-c	0
4	17.00	$\frac{1}{2}$	$-\frac{1}{2}$	-1	$-\frac{1}{2}$	$\frac{1}{2}$	1
5	15.55	d	-b	-c	$\frac{1}{2}$	a	0
6	13.99	0	-1	0	1	0	-1
7	12.45	-d	-b	c	$\frac{1}{2}$	-a	0
8	11.19	$-\frac{1}{2}$	$-\frac{1}{2}$	1	$-\frac{1}{2}$	$-\frac{1}{2}$	1
9	10.01	-c	0	c	-1	c	0
10	9.02	-b	$\frac{1}{2}$	0	$-\frac{1}{2}$	b	-1
11	8.29	-a	b	-c	$\frac{1}{2}$	-d	0
12	4.00	-1	1	-1	1	-1	1
	166.68	33.567	-1.3485	-0.5640	-0.775	+0.244	0.130
a_q	6.945	2.797	-0.112	-0.047	-0.065	0.020	0.011

j	G_j	sin x	sin 2x	sin 3x	sin 4x	sin 5x	sin 6x
0	0	0	0	0	0	0	0
1	-0.07	d	$\frac{1}{2}$	c	b	a	1
2	-0.10	$\frac{1}{2}$	b	1	b	$\frac{1}{2}$	0
3	-0.11	c	1	c	0	-c	-1
4	0.16	b	b	0	-b	-b	0
5	0.65	a	$\frac{1}{2}$	-c	-b	d	1
6	1.19	1	0	-1	0	1	0
7	1.55	a	$-\frac{1}{2}$	-c	b	d	-1
8	1.49	b	-b	0	b	-b	0
9	1.11	c	-1	c	0	-c	1
10	0.58	$\frac{1}{2}$	-b	1	-b	$\frac{1}{2}$	0
11	0.13	d	$-\frac{1}{2}$	c	-b	a	-1
12	0	0	0	0	0	0	0
		5.7066	-0.178	-1.5161	1.1691	-0.338	0.120
b_q		0.476	-0.015	-0.126	0.097	-0.028	0.010

$$f(x) = 6.945 + 2.797 \cos x - 0.112 \cos 2x - 0.047 \cos 3x$$
$$- 0.065 \cos 4x + 0.020 \cos 5x + 0.011 \cos 6x$$
$$+ 0.476 \sin x - 0.015 \sin 2x - 0.126 \sin 3x$$
$$+ 0.097 \sin 4x - 0.028 \sin 5x + 0.010 \sin 6x.$$

Exercises

1. Express in terms of sines:

j:	0	1	2	3	4	5	6	7	8	9	10	11	12
y_j:	0	1	2	3	4	4	4	4	4	5	2	1	0

2. Express in terms of cosines:

j:	0	1	2	3	4	5	6	7	8	9	10	11	12
y_j:	5	5	4	3	2	1	0	-1	-2	-3	-4	-5	-5

3. Express in terms of sines and cosines:

j:	0	1	2	3	4	5	6	7	8	9	10	11	12
y_j:	-2	-2	-1	0	1	2	3	4	4	2	0	-2	-2

81. THE GRAM-CHARLIER APPROXIMATION

The principal applications of the method of approximation about to be described are for functions which approach zero very rapidly as $|x|$ becomes infinite. A large class of functions representing frequency distributions have this property.

Let

$$\varphi_0(x) = \frac{1}{\sqrt{2\pi}} e^{-x^2/2}, \tag{1}$$

and let the successive derivatives of $\varphi_0(x)$ with respect to x be denoted by $\varphi_1(x)$, $\varphi_2(x)$, $\varphi_3(x)$, ... Upon performing the differentiations it is found that

$$\varphi_1(x) = -x\,\varphi_0(x),$$
$$\varphi_2(x) = (x^2 - 1)\,\varphi_0(x),$$
$$\varphi_3(x) = -(x^3 - 3x)\,\varphi_0(x),$$
$$\varphi_4(x) = (x^4 - 6x^2 + 3)\,\varphi_0(x),$$

$$\varphi_5(x) = (-x^5 + 10x^3 - 15x)\ \varphi_0(x),$$
$$\varphi_6(x) = (x^6 - 15x^4 + 45x^2 - 15)\ \varphi_0(x),$$
$$\varphi_7(x) = (-x^7 + 21\,x^5 - 105x^3 + 105x)\ \varphi_0(x),$$
$$\varphi_8(x) = (x^8 - 28x^6 + 210x^4 - 420x^2 + 105)\ \varphi_0(x),$$

and in general

$$\varphi_n(x) = H_n(x)\ \frac{1}{\sqrt{2\pi}}\ e^{-x^2/2}. \tag{2}$$

The function $H_n(x)$ is a polynomial of degree n, even if n is even, odd if n is odd. It is called an "Hermitian" polynomial in honor of Hermite.

Differentiation of (2) gives

$$\varphi_{n+1}(x) = H_{n+1}(x)\ \varphi_0(x) = \left[H_n'(x) - xH_n(x)\right]\ \varphi_0(x)$$

so that the H's satisfy the identity

$$H_{n+1}(x) = H_n'(x) - xH_n(x). \tag{3}$$

From the first few H's we observe that

$$H_1'(x) = -1 = -H_0(x)$$
$$H_2'(x) = 2x = -2H_1(x)$$
$$H_3'(x) = -3x^2 + 3 = -3H_2(x)$$
etc.

suggesting that in general

$$H_n'(x) = -nH_{n-1}(x). \tag{4}$$

That (4) is in fact true may be proved by induction by use of the relation (3). The proof is left to the reader.

We next investigate the integral

$$\int_{-\infty}^{\infty} H_n(x)\ \varphi_m(x)\ dx. \tag{5}$$

It is evident from the definitions that

$$\int_{-\infty}^{\infty} H_n(x)\,\varphi_m(x)\ dx = \int_{-\infty}^{\infty} H_m(x)\,\varphi_n(x)\ dx$$

so that without loss of generality we may treat only the case where $n \leqq m$ in (5). Integrating (5) by parts and noting that the integrated terms vanish at the limits because of the exponential in $\varphi_{m-1}(x)$ we arrive at the equation

$$\int_{-\infty}^{\infty} H_n(x)\,\varphi_m(x)\ dx = n \int_{-\infty}^{\infty} H_{n-1}(x)\ \varphi_{m-1}(x)\ dx$$

and finally by repetition of the process

$$\int_{-\infty}^{\infty} H_n(x)\ \varphi_m(x)\ dx = n! \int_{-\infty}^{\infty} H_o(x)\ \varphi_{m-n}(x)\ dx,$$

since $n \leqq m$. If $m = n$ the value of the integral on the right is 1, otherwise 0. Hence

$$\int_{-\infty}^{\infty} H_n(x)\ \varphi_m(x)\ dx = \begin{cases} 0 & \text{if } m \neq n, \\ n' & \text{if } m = n. \end{cases} \tag{6}$$

This may be written

$$\int_{-\infty}^{\infty} \varphi_o(x)\ H_n(x)\ H_m(x)\ dx = \begin{cases} 0 & \text{if } m \neq n, \\ n! & \text{if } m = n. \end{cases} \tag{7}$$

Hence the polynomial $H_n(x)$ form an orthogonal set for the interval $-\infty$ to ∞ with respect to the weight function $\varphi_o(x)$. These functions are therefore suitable for weighted Least Square approximations over the infinite interval. Three types of approximation will be mentioned.

A. If we wish to represent a function f(x) by a polynomial in such a way as to emphasize accuracy near the origin at the expense of accuracy for |x| large, we may minimize

$$E = \int_{-\infty}^{\infty} \varphi_o(x) \left[f(x) - \sum_{i=0}^{n} a_i H_i(x) \right]^2 dx,$$

where it is implied that f(x) is such that the infinite integral will converge. Upon evaluating the coefficients a_k in the usual manner we get

$$a_k = \frac{1}{k!} \int_{-\infty}^{\infty} f(x)\ \varphi_k(x)\ dx.$$

B. If we wish to approximate f(x) by a function of the form $y = \sum_{i=0}^{n} b_i H_i(x) \sqrt{\varphi_o(x)}$, we may minimize

$$E = \int_{-\infty}^{\infty} \left[f(x) - \sum_{i=0}^{n} b_i H_i(x) \sqrt{\varphi_o(x)} \right]^2 dx.$$

Here is it found that

$$b_k = \frac{1}{k!} \int_{-\infty}^{\infty} f(x)\ H_k(x) \sqrt{\varphi_o(x)}\ dx$$

again with the assumption that the integrals converge.

C. If we approximate f(x) by a function of the form

$$(8) \qquad y = \sum_{i=0}^{n} c_i \varphi_i(x)$$

with the weight function $\varphi_o^{-1}(x)$ we minimize

$$E = \int_{-\infty}^{\infty} \varphi_0^{-1}(x) \left[f(x) - \sum_{i=0}^{n} c_i \varphi_i(x) \right]^2 dx.$$

Here it turns out that

$$c_k = \frac{1}{k!} \int_{-\infty}^{\infty} f(x) H_k(x)\, dx. \tag{9}$$

Since the $H_k(x)$ are polynomials it is evident that the infinite integrals above will converge only if $f(x)$ approaches zero rapidly as $|x|$ becomes infinite.

We limit our discussion to case C, which is the case most frequently used. It can be shown that if the coefficients in (8) are determined by equations (9) and if n is sufficiently large, then y will be a good approximation to $f(x)$ for a very general class of functions $f(x)$ provided only that $f(x)$ approaches zero with sufficient rapidity as $|x|$ becomes infinite. However, in practical applications in order to secure a good approximation without making n needlessly large, it is important first of all to choose the origin and the scale for the variable x in $f(x)$ in such a way that

$$\int_{-\infty}^{\infty} x f(x)\, dx = 0, \quad \text{and} \quad \int_{-\infty}^{\infty} x^2 f(x)\, dx = 1.$$

If the original function is $g(s)$ with moments m_q, where

$$m_q = \int_{-\infty}^{\infty} s^q g(s)\, ds, \quad q = 0, 1, 2, \ldots$$

and if $g(s)$ does not change sign, the transformation

$$s = ax + b, \qquad g(s) = cf(x)$$

where

$$a^2 = \frac{m_0 m_2 - m_1^2}{m_0^2}, \quad b = \frac{m_1}{m_0}, \quad c = \frac{m_0}{a},$$

gives a function f(x) with the moments

$$(10) \qquad \mu_q = \int_{-\infty}^{\infty} x^q f(x)\, dx$$

where $\mu_o = 1$, $\mu_1 = 0$, $\mu_2 = 1$. It will be assumed that this transformation has been made.

Then by putting the actual polynomial expression for $H_k(x)$ in (9) and using (10) we get

$$c_o = 1, \qquad c_1 = 0, \qquad c_2 = 0,$$

$$c_3 = -\mu_3/6, \qquad c_4 = (\mu_4 - 3)/24,$$

$$c_5 = (-\mu_5 + 10\,\mu_3)/120, \quad c_6 = (\mu_6 - 15\,\mu_4 + 30)/720,$$

etc.

In this way the coefficients c_k are all expressible in terms of the moments.

The function G(s) introduced in Article 30 in connection with error terms is a function which is identically zero outside of a given interval, and is therefore a type of function to which the present method of approximation is adapted. If in formula (5) of Article 30 we let

$$f(x) = \frac{q!}{(n+1+q)!} x^{n+1+q}$$

we have

$$\frac{q!}{(n+1+q)!} R\left[x^{n+1+q}\right] = \int_{-\infty}^{\infty} s^q G(s)\, ds.$$

Hence, the qth moment of $G(s)$ is given by

$$m_q = \frac{q!}{(n + 1 + q)!} R \left[x^{n+1+q}\right].$$

As a particular example, we consider Simpson's Rule formed for the three points $-h$, 0, h, where $n = 4$ and

$$R(f) = f(h) - f(-h) - \frac{h}{3}\left[f'(h) + 4f'(0) + f'(-h)\right].$$

Using this formula we obtain for the moments

$$m_q = 0 \qquad \text{if } q \text{ is odd}$$

$$m_q = -\frac{2q!\,(q + 2)}{3(5 + q)!} h^{5+q} \qquad \text{if } q \text{ is even.}$$

Whence $m_0 = -\frac{h^5}{90}$, $\quad m_2 = \frac{h^7}{945}$.

$$m_4 = -\frac{h^9}{3789}, \qquad m_6 = -\frac{h^{11}}{10395}, \qquad m_8 = -\frac{h^{13}}{23166}.$$

After the transformation

$$G(s) = -\frac{h^5}{90} f(x),$$

$$s = \sqrt{\frac{2}{21}}\, hx,$$

we have for the moments of $f(x)$

$\mu_q = 0$ if q is odd, and

$$\mu_0 = 1, \quad \mu_2 = 1, \quad \mu_4 = \frac{21}{8}, \quad \mu_6 = \frac{441}{44}, \quad \mu_8 = \frac{324135}{6864}.$$

From these we find the coefficients c_k:

$c_k = 0 \qquad$ if k is odd,

$$c_0 = 1, \quad c_2 = 0, \quad c_4 = -\frac{1}{64}, \quad c_6 = \frac{57}{63360}, \quad c_8 = \frac{103}{1464320}.$$

Hence, approximately

$$G(s) = -\frac{h^5}{90}\left[\varphi_0(x) - \frac{\varphi_4(x)}{64} + \frac{57\,\varphi_6(x)}{63360} + \frac{103\,\varphi_8(x)}{1464320}\right]$$

where $x = \sqrt{\frac{21}{2}}\,\frac{s}{h}$. In the table below are given the values of the function $g(s) = -90h^{-5}G(s)$, and the values as computed from the approximation, together with the differences.

$$c_0 = 1, \quad c_4 = -0.01562, \quad c_6 = 0.00090, \quad c_8 = 0.00007$$

x	g(s)	$\varphi_0(x)$	$\varphi_4(x)$	$\varphi_6(x)$	$\varphi_8(x)$	Approx.	Diff.
0	0.3858	0.3989	1.1968	-5.984	41.9	0.3778	0.0080
0.5	0.3413	0.3521	0.5501	-1.645	4.5	0.3423	-0.0010
1.0	0.2455	0.2420	-0.4839	3.872	-31.9	0.2508	-0.0053
1.5	0.1428	0.1295	-0.7043	2.811	-9.1	0.1424	0.0004
2.0	0.0617	0.0540	-0.2700	-0.594	13.4	0.0586	0.0031
2.5	0.0153	0.0175	0.0800	-1.324	6.5	0.0157	-0.0004
3.24	0	0.0210	0.1053	-0.080	-3.0	0.0002	-0.0002

Exercises

Obtain a few terms of the Gram-Charlier Approximation to the following functions, where

$$\overline{(x-a)}^n = \begin{cases} (x-a)^n & \text{if } x > a \\ 0 & \text{if } x < a. \end{cases}$$

1. $f(x) = (\overline{x-3})^2 - 3(\overline{x-1})^2 + 3(\overline{x+1})^2 - (\overline{x+3})^2.$
2. $f(x) = (\overline{x-4})^3 - 4(\overline{x-2})^3 + 6(\overline{x})^3 - 4(\overline{x+2})^3 + (\overline{x+4})^3.$
3. $f(x) = (\overline{x-5})^4 - 5(\overline{x-3})^4 + 10(\overline{x-1})^4 - 10(\overline{x+1})^4 + 5(\overline{x+3})^4 - (\overline{x+5})^4.$

(Except for a constant factor these are special cases of the function

$$f_n(x) = \frac{1}{2^n(n-1)!}\sum_{k=0}^{n}(-1)^k\binom{n}{k}(\overline{x-n+2k})^{n-1}.$$

This is the frequency function for the error of the sum

$$u_1 + u_2 + \cdot \cdot \cdot + u_n$$

where each u is subject to an error not greater than unity, all errors between -1 and 1 being equally likely.)

82. CASE OF EQUALLY SPACED POINTS

In analogy with the development in Article 81 for continuous approximation, let us now consider the problem of representing a frequency function by means of values at equally spaced points. For this purpose we employ a set of functions $H_t^{(n)}(s)$ defined by

$$(1) \qquad H_t^{(n)}(s) = \Delta^t_s \binom{n-t}{s-t},$$

for n, s, and t integers. In particular

$$H_0^{(n)}(s) = \binom{n}{s}$$

$$H_1^{(n)}(s) = \Delta_s \binom{n-1}{s-1} = \binom{n-1}{s} - \binom{n-1}{s-1}$$

$$= \frac{(n-1)!}{s!\,(n-s-1)!} - \frac{(n-1)!}{(s-1)!\,(n-s)!}$$

$$= \frac{n!}{s!\,(n-s)!}\left[\frac{n-s}{n} - \frac{s}{n}\right]$$

$$= \binom{n}{s}\left(1 - \frac{2s}{n}\right).$$

Similarly

$$H_2^{(n)}(s) = \binom{n}{s}\left(1 - \frac{4s}{n} + 4\frac{s(s-1)}{n(n-1)}\right)$$

and in general

$$(2) \qquad H_t^{(n)}(s) = \binom{n}{s}\sum_{k=0}^{t}(-1)^k 2^k \frac{\binom{s}{k}\binom{t}{k}}{\binom{n}{k}}.$$

We note that the quantity

$$\binom{n}{t} H_t^{(n)}(s) = \binom{n}{t}\binom{n}{s}\sum_{k=0}^{t}(-1)^k 2^k \frac{\binom{s}{k}\binom{t}{k}}{\binom{n}{k}}$$

is actually symmetrical with respect to s and t, since the sum on the right always terminates with k equal to the <u>smaller</u> of the two integers s and t. Hence

$$(3) \qquad \binom{n}{t} H_t^{(n)}(s) = \binom{n}{s} H_s^{(n)}(t).$$

The functions $H_t^{(n)}(s)$ satisfy the fundamental identity

$$(4) \qquad H_{t+1}(s+1) = H_t(s+1) - H_{t+1}(s) - H_t(s).$$

For $\quad H_t(s+1) - H_t(s) = \Delta_s H_t(s)$

$$= \Delta_s^{t+1}\binom{n-t}{s-t}.$$

But $\qquad H_{t+1}(s) = \Delta_s^{t+1}\binom{n-t-1}{s-t-1}$

Hence the right-hand member of (4) is

$$\Delta_s^{t+1} \left[\binom{n-t}{s-t} - \binom{n-t-1}{s-t-1}\right]$$

$$= \Delta_s^{t+1} \binom{n-t-1}{s-t} = H_{t+1}^{(n)}(s+1).$$

For any given value of n equation (4) furnishes a simple way to prepare a numerical table of $H_t^{(n)}(s)$ for all integral values of t and s. This will be a two way table in which the row for $t = 0$ has the value $\binom{n}{s}$ for each s, while the column for $s = 0$ has the value 1 for each t. When these entries are placed in the first row and first column of the table all the remaining entries are rapidly computed by means of (4). For example the completed table in the case $n = 6$ appears as follows:

t \ s	0	1	2	3	4	5	6
0	1	6	15	20	15	6	1
1	1	4	5	0	-5	-4	-1
2	1	2	-1	-4	-1	2	1
3	1	0	-3	0	3	0	-1
4	1	-2	-1	4	-1	-2	1
5	1	-4	5	0	-5	4	-1
6	1	-6	15	-20	15	-6	1

The numbers in the table have interesting properties. We note that the sum of products of corresponding elements in the i-th row and j^{th} column is zero if $i \neq j$ and is 2^6 if $i = j$. This is a special case of the general theorem

$$(5) \qquad \sum_{k=0}^{n} H_k^{(n)}(s)\, H_t^{(n)}(k) = \begin{cases} 0 & \text{if} \quad s \neq t \\ 2^n & \text{if} \quad s = t. \end{cases}$$

The orthogonal property (5) shows that the functions $H_t^{(n)}(s)$ are suited to approximations by means of Least Squares. Let $f(x)$ be a function with values given for $x = 0, 1, 2, \ldots, n$, and equal to zero for $x \leqq -1$ or $x \geqq n+1$. Let us determine the constants c_k so as to minimize the sum

$$(6) \qquad E = \sum_{x=0}^{n} \frac{1}{\binom{n}{x}} \left[f(x) - \sum_{k=0}^{m} c_k\, H_k^{(n)}(x) \right]^2, \qquad m \leqq n.$$

If we differentiate E with respect to c_q, multiply by $-\frac{1}{2}\binom{n}{q}$, and equate the result to zero we obtain

$$\sum_{k=0}^{m} c_k \sum_{x=0}^{n} H_k^{(n)}(x)\, H_q^{(n)}(x) \binom{n}{q} \Big/ \binom{n}{x}$$

$$= \sum_{x=0}^{n} f(x)\, H_q^{(n)}(x) \binom{n}{q} \Big/ \binom{n}{x}.$$

But by (3) this reduces to

$$\sum_{k=0}^{m} c_k \sum_{x=0}^{n} H_k^{(n)}(x)\, H_x^{(n)}(k) = \sum_{x=0}^{n} f(x)\, H_x(q)$$

and by (5) we get finally

$$c_q = \frac{1}{2^n} \sum_{s=0}^{n} f(s)\, H_s(q).$$

If in (6) we take $m = n$, the resulting approximation

$$y(x) = \sum_{k=0}^{m} c_k H_k^{(n)}(x)$$

coincides with f(x) exactly at the n + 1 points x = 0, 1, . . ., n, and y(x) in this case furnishes an interpolation formula for f(x). If, however, m is less than n, y(x) will not in general coincide with f(x) at the given points. The values obtained for y(x) at the points x = 0, 1, . . ., n are in this case a smoothed set of values which in view of the Fundamental Theorem will under suitable conditions be closer to the hypothetical "true" values than are the original values of f(x).

Example: The scores of 1880 students on the A.C.E. Test were distributed as follows:

	Score		Number of Students
x	From	To	f(x)
0	1	20	0
1	21	40	3
2	41	60	56
3	61	80	259
4	81	100	525
5	101	120	654
6	121	140	299
7	141	160	75
8	161	180	9
9	181	200	0

For this case n = 9. Let us arbitrarily assume that m = 4. Then the complete computation of the smoothed values appears below:

k	f(k)	x = 0	1	2	3	4	5	6	7	8	9	y(x)
0	0	1	9	36	84	126	126	84	36	9	1	1.6
1	3	1	7	20	28	14	-14	-28	-20	-7	-1	2.2
2	56	1	5	8	0	-14	-14	0	8	5	1	47.5
3	259	1	3	0	-8	-6	6	8	0	-3	-1	260.8
4	525	1	1	-4	-4	6	6	-4	-4	1	1	562.2
5	654	1	-1	-4	4	6						598.9
6	299	1	-3	0	8	-6						323.8
7	75	1	-5	8	0	-14						77.6
8	9	1	-7	20	-28	14						5.0
9	0	1	-9	36	-84	126						0.4
	C_x =	1880	-386	-3428	668	2060	÷ 512					

Starting with $n = 9$, $m = 4$, we use (4) to construct as much of the table of values of $H_k(x)$ as is needed. For $x = 0, 1, 2, 3, 4$, we calculate the sum of products

$$C_x = \sum_{k=0}^{9} f(k)\, H_k(x)$$

obtaining in this way $C_o = 1880$, $C_1 = -386$, etc., which are entered at the foot of the appropriate columns. Finally, to find $y(x)$ we calculate the sums of products

$$\sum_{q=0}^{4} C_q\, H_q(x)$$

for $x = 0, 1, 2, \ldots, 9$, divide each by $2^9 = 512$, and record the result under $y(x)$ in the right-hand column.

The foregoing procedure provides an effective method for smoothing data obtained for frequency distributions with equal intervals provided

a) the values tend to zero at both ends,

b) the number $n + 1$ of subdivisions is not too great.

Exercises

1. Construct a table of $H_k^{(n)}(x)$ for $n = 5$; $n = 12$;

n = 20.

2. In the example of Article 73, group the data of successive pairs of entries, so as to cut the number of entries in half. Then apply this method of smoothing.

CHAPTER XI
SIMPLE DIFFERENCE EQUATIONS

An equation involving an independent variable x, a dependent variable y, and one or more derivatives $\frac{dy}{dx}$, $\frac{d^2y}{dx^2}$, etc., is a differential equation. Similarly, an equation involving an independent variable x, a dependent variable y, and one or more differences Δy, $\Delta^2 y$, etc., is a difference equation. There is consequently a close analogy between differential equations and difference equations, and it would appear that much of the theory of differential equations could be carried over bodily to difference equations. The actual situation, however, is not that simple, and in fact the general solution of difference equations encounters difficulties not anticipated from our experience with differential equations. In order to keep the analysis at an elementary level, it is necessary to restrict our discussion of difference equations to a small number of particularly simple cases.

83. SOLUTION OF A DIFFERENCE EQUATION

It is convenient to make a change of the independent variable x with $\Delta x = h$ to a new variable s where

$$x = hs$$

so that $\Delta s = 1$. From now on we shall suppose that this has been done. Also since $\Delta u_s = u_{s+1} - u_s$, $\Delta^2 u_s = u_{s+2} - 2u_{s+1} + u_s$, etc., we shall frequently write a difference equation of first order in terms of s, u_s, u_{s+1}, rather than in terms of s, u_s, Δu_s, and similarly for higher orders. Thus

$$u_{s+1} = su_s - s^2$$

is a difference equation of first order,

$$u_{s+2} - su_{s+1} + u_s = 0$$

is a difference equation of second order, etc.

The fundamental question is, "What is meant by a <u>solution</u> of a difference equation?" We shall try to clarify this question by means of an example. Consider the equation

$$u_{s+1} = 2u_s + s^2 - 2s - 1 \tag{1}$$

together with the initial value $u_0 = 1$.

First let us treat the equation by purely numerical steps. Setting $s = 0$, we have, since $u_0 = 1$,

$$u_1 = 2 - 1 = 1$$

and when $s = 1$

$$u_2 = 0.$$

Proceeding thus step by step we get a table of values:

s:	0	1	2	3	4.	5	6	7	8	9	10	etc.
u_s:	1	1	0	-1	0	7	28	79	192	431	924	

This table of values constitutes a particular numerical solution of the difference equation. It is not a continuous function of s, but is defined only for integral values of s. We might call it a <u>particular discrete solution</u> of the difference equations. Evidently, this solution is uniquely determined. Now the continuous function of s

$$u_s = C2^s - s^2 \tag{2}$$

also satisfies the difference equation, since

$$u_{s+1} = C2^{s+1} - (s + 1)^2$$
$$= 2C2^s - s^2 - 2s - 1$$

whence

$$u_{s+1} = 2u_s + s^2 - 2s - 1$$

If C is determined to satisfy the condition $u_s = 1$ when $s = 0$ we find that $C = 1$ and

$$u_s = 2^s - s^2$$

is a <u>particular</u> <u>continuous</u> <u>solution</u> of the difference equation. For integral values of s this solution yields the numerical values already obtained.

Since the solution (2) contains an arbitrary constant we would be tempted, in analogy with differential equations, to suppose that (2) represents the general solution of (1) and that all particular solutions can be found by giving C suitable particular values. In a limited sense this is true. All <u>discrete</u> <u>particular</u> <u>solutions</u> of (1) are contained in (2). But nevertheless (2) is not the general solution of (1) since

$$u_s = 2^s(a + b \cos 2\pi s + c \sin 2\pi s) - s^2 \tag{3}$$

for example, where a, b, c are constants, is also a solution of (1), as the reader may verify by substitution, using the periodic property of the sine and cosine. In fact, having tested (3), we now see that

$$u_s = 2^s w(s) - s^2 \tag{4}$$

is a solution of (1), where $w(s)$ is <u>any</u> periodic function with period 1. Equation (4), rather than (2), gives the <u>general</u> <u>solution</u> of (1). Generalizing from this example we shall state (without proof) that the general solution of a first-order difference equation contains an arbitrary periodic function with period unity.

Conversely, suppose that

$$G\big(s,\ u_s,\ w(s)\big) = 0$$

is an equation containing s, u_s, and a periodic function $w(s)$ with period 1. We have, after replacing s by $s + 1$,

$$G\big(s + 1,\ u_{s+1},\ w(s)\big) = 0$$

Upon eliminating $w(s)$ between these two equations we obtain a relation

$$F(s,\ u_s,\ u_{s+1}) = 0$$

connecting s, u_s, and u_{s+1}, and this is a difference equation of order one. If we start with a relation containing two periodic functions, the elimination leads to a second-order difference equation, and so on.

To review the discussion we see that when a difference equation is presented to us for solution, it is important that we know what sort of solution is desired. If what we want is a particular discrete solution whose values are found for a set of values of s proceeding at unit intervals, such a solution can usually be obtained with no other difficulty than the computational labor involved. If on the other hand, we want a continuous solution expressed in closed form in terms of known elementary functions, the situation is very different. Such solutions are obtainable only for a relatively small class of difference equations.

Exercises

Obtain six values of the discrete solution which satisfies the difference equation and the given condition.

1. $u_{s+1} - u_s = s$, $u_0 = 1$
2. $u_{s+1} = su_s$, $u_1 = 1$
3. $u_{s+1} = u_s/s$, $u_1 = 1$
4. $u_{s+1} = u_s^2$, $u_0 = 1$
5. $u_{s+1} - 2u_s = 2^{s+1}$ $u_0 = 0$
6. $su_{s+1} - u_s = 0$ $u_1 = 1$
7. $3u_{s+2} - 5u_{s+1} + 3u_s = 0$ $u_0 = 0$, $u_1 = 1$.
8. $u_{s+2} - 2u_{s+1} + u_s = 0$ $u_0 = 4$, $u_1 = 5$.
9. $u_{s+2} - 6u_{s+1} + u_s = 0$ $u_0 = 0$, $u_1 = 1$.
10. $u_{s+2} + 3u_{s+1} + 2u_s = 0$ $u_0 = 1$, $u_1 = 0$.

84. CALCULUS OF DIFFERENCES

In analogy with the simple differential equation $dy/dx = f(x)$, the general solution of which is the indefinite integral of $f(x)$, we may consider the difference equation

$$u_{s+1} - u_s = f(s). \tag{1}$$

Just as in beginning calculus the indefinite integral is developed as the inverse of a differential and methods of integrating particular functions are worked out from results previously established by differentiation, so here we can solve a number of problems by finding first of all the difference formulas for various elementary functions. A partial list of elementary difference formulas is supplied below. The reader should verify each of these.

List of Differences

(In this list w, w_1, w_2, etc., denote periodic functions of period 1.)

1. $\Delta w = 0.$

2. $\Delta w \cdot u = w \cdot \Delta u.$

3. $\Delta(w_1 . u + w_2 . v) = w_1 . \Delta u + w_2 . \Delta v.$

4. $\Delta u_s v_s = v_{s+1} \Delta u_s + u_s \Delta v_s$

   $= u_{s+1} \Delta v_s + v_s \Delta u_s.$

5. $\Delta \frac{u_s}{v_s} = \frac{v_s \Delta u_s - u_s \Delta v_s}{v_s v_{s+1}}.$

6. $\Delta s^{(n)} = ns^{(n-1)}.$

(Since any polynomial can be expressed as $a_0 s^{(n)} + a_1 s^{(n-1)} + \ldots + a_n$ the difference of any polynomial is obtainable from 6. The difference of any rational fraction can be found by 5 and 6.)

7. $\Delta \frac{1}{s^{(n)}} \doteq \frac{-n}{(s + 1)^{(n+1)}}.$

8. $\Delta a^s = a^s(a - 1).$

9. $\Delta a^{-s} = a^{-s}(\frac{1}{a} - 1), \; a \neq 0.$

10. $\Delta \ln s = \ln(1 + \frac{1}{s}), \; s > 0.$

11. $\Delta \sin ms = 2 \sin \frac{m}{2} \cos (ms + \frac{m}{2}).$

12. $\Delta \cos ms = -2 \sin \frac{m}{2} \sin (ms + \frac{m}{2}).$

13. $\Delta \sinh ms = 2 \sinh \frac{m}{2} \cosh (ms + \frac{m}{2}).$

14. $\Delta \cosh ms = 2 \sinh \frac{m}{2} \sinh (ms + \frac{m}{2}).$

15. $\Delta (u_{s-1} + u_{s-2} + \ldots + u_{s-n}) = u_s - u_{s-n}.$

16. $\Delta (u_{s-1} \cdot u_{s-2} \cdot \ldots \cdot u_{s-n}) = u_{s-1} \cdot u_{s-2} \cdot \ldots \cdot u_{s-n+1}$

    $(u_s - u_{s-n}).$

The remaining formulas involve the gamma function $\Gamma(x)$, with which the reader is assumed to be acquainted. The properties of $\Gamma(x)$ which we need here are:

The recurrence relation

$$\Gamma(x+1) = x\,\Gamma(x) \tag{1}$$

For integral values of x, say $x = n \geq 0$

$$\Gamma(n+1) = n! \tag{2}$$

The notation $\Gamma(x+1) = x!$ is frequently used even when x is not integral.

The asymptotic expansion

$$\Gamma(x+1) = \sqrt{2\pi x}\, e^{-x} x^{x}\left[1 + \frac{1}{12x} + \frac{1}{288x^{2}} - \frac{139}{51840x^{3}} - \frac{571}{2488320x^{4}} + \cdots\right] \tag{3}$$

The logarithmic derivative of $\Gamma(x+1)$ is called the digamma function and is denoted here by $\Psi(x)$ so that

$$\Psi(x) = \frac{d}{dx}\ln \Gamma(x+1). \tag{4}$$

The digamma function has the asymptotic expansion

$$\Psi(x) = \frac{1}{2}\ln x(x+1) + \frac{1}{6}\left(\frac{1}{x} - \frac{1}{x+1}\right) - \frac{1}{90}\left(\frac{1}{x^{3}} - \frac{1}{(x+1)^{3}}\right) + \frac{1}{210}\left(\frac{1}{x^{5}} - \frac{1}{(x+1)^{5}}\right) - \cdots . \tag{5}$$

The derivatives $\Psi'(x)$, $\Psi''(x)$, $\Psi'''(x)$, etc. are called the trigamma, tetragamma, pentagamma, etc., functions respectively. Their asymptotic expansions are obtained by differentiating (5).

Values of $\Gamma(x+1)$ and of $\Psi(x)$ are given in Table VIII. Use of the tables and asymptotic formulas will be illustrated in subsequent examples.

<u>Difference Formulas, Continued</u>

17. $\Delta\Gamma(s+1) = s\,\Gamma(s+1)$

18. $\Delta \ln \Gamma(s+1) = \ln(s+1)$

19. $$\frac{\Gamma(s+1+a_1)\quad\Gamma(s+1+a_2)}{\Gamma(s+1+b_1)\quad\Gamma(s+1+b_2)} = \frac{(s+a_1)(s+a_2)}{(s+b_1)(s+b_2)}\,\frac{\Gamma(s+a_1)\quad\Gamma(s+a_2)}{\Gamma(s+b_1)\quad\Gamma(s+b_2)}$$

20. $\Gamma(s+1)\,\Gamma(s)\quad\Gamma(s-1)\;.\;.\;.\quad\Gamma(s-n+2)$

$$= s^{(n)}\quad\Gamma(s)\quad\Gamma(s-1)\;\ldots\;\Gamma(s-n+1).$$

21. $a^{s+1}\quad\Gamma(s+s+\frac{b}{a}) = (as+b)a^s\quad\Gamma(s+\frac{b}{a}).$

22. $\Delta\Psi(s) = \dfrac{1}{(s+1)}.$

23. $\Delta\Psi'(s) = \dfrac{-1}{(s+1)^2}.$

24. $\Delta\Psi''(s) = \dfrac{2}{(s+1)^3}.$

25. $\Delta\Psi'''(s) = \dfrac{-6}{(s+1)^4}.$

26. $$\Delta\left\{\frac{1}{2ai}\left[\Psi(s+ai) - \Psi(s-ai)\right]\right\} = \frac{-1}{(s+1)^2+a^2},$$

where $i = \sqrt{-1}$.

85. THE DIFFERENCE EQUATION $\Delta u_s = f(s)$.

The difference formulas of Article 84 provide the solution for a number of difference equations in the form

$$\Delta u_s = f(s).$$

Case I. Polynomial. If $f(s)$ is a polynomial expressed (as is always possible) in terms of factorials

$$f(s) = a_0 s^{(n)} + a_1 s^{(n-1)} + \dots + a_n$$

it follows from difference formula 6 that

$$u_s = \frac{a_0 s^{(n+1)}}{n+1} + \frac{a_1 s^{(n)}}{n} + \dots + a_n s + w.$$

Example 1. Find the sum of the cubes of the first 100 integers.

Here the appropriate difference equation is

$$\Delta u_s = s^3 = s^{(3)} + 3s^{(2)} + s^{(1)}$$

and the solution is

$$u_s = \frac{s^{(4)}}{4} + s^{(3)} + \frac{s^{(2)}}{2} + w(s).$$

From difference formula 15 it is clear that the desired sum is equal to

$$u_{101} - u_0 = 3232000.$$

Case II. Rational Fractions. If $f(s)$ is any rational fraction, we first of all reduce it to a proper fraction by division if necessary. The integral part, if any, is handled by Case I. The degree of the numerator is now less than the denominator, and the fraction can be expressed in the usual manner by partial fractions. If we first suppose that the linear factors of the denominator are all real and distinct, we have a difference equation of the type

$$\Delta u_s = \frac{A}{s+a} + \frac{B}{s+b} + \dots + \frac{K}{s+k}.$$

From the difference formula 22 it is now evident that

$$u_s = A\Psi(s+a-1) + B\Psi(s+b-1) + \dots + K\Psi(s+k-1) + w(s).$$

In this connection it is helpful to realize that the function $\Psi(s)$ plays a role in difference calculus analogous to that of $\ln(1 + s)$ in infinitesimal calculus. In fact the graph of $\Psi(s)$ is quite similar in appearance to that of $\ln(s + 1)$ for $s > -1$.

Example 2. Find the value of

$$\sum_{k=1}^{50} \frac{1}{(k+1)(2k+1)}.$$

Here the difference equation is

$$\Delta u_s = \frac{1}{(s+1)(2s+1)} = \frac{1}{s+\frac{1}{2}} - \frac{1}{s+1}$$

whence

$$u_s = \Psi(s - 0.5) - \Psi(s)$$

and the desired sum is

$$u_{51} - u_1 = \Psi(50.5) - \Psi(51) - \Psi(0.5) + \Psi(1).$$

From Table VIII

$$\Psi(1) = 0.42278$$
$$\Psi(0.5) = 0.03649.$$

From the asymptotic formula

$$\Psi(50.5) - \Psi(51) = \frac{1}{2}\ln\left[\frac{(50.5)(51.5)}{51 \cdot 52}\right] + \frac{1}{6}\left[\frac{1}{50.5} - \frac{1}{51} - \frac{1}{51.5} + \frac{1}{52} + \dots\right] = -0.00976.$$

Hence

$$\sum_{k=1}^{50} \frac{1}{(k+1)(2k+1)} = 0.37653.$$

Example 3. Find the sum of the convergent infinite series

$$\sum_{k=1}^{\infty} \frac{1}{(k+1)(2k+1)(5k+2)}.$$

The difference equation will be

$$\Delta u_s = \frac{1}{(s+1)(2s+1)(5s+2)}$$

$$= \frac{1}{3}\left[\frac{1}{s+1} - \frac{6}{s+1/2} + \frac{5}{s+2/5}\right],$$

and its solution is

$$u_s = \frac{1}{3}\left[\Psi(s) - 6\Psi(s-\tfrac{1}{2}) + 5\Psi(s-\tfrac{3}{5})\right] + w(s).$$

The sum of n terms of the series will be

$$u_{n+1} - u_1 = \frac{1}{3}\left[\Psi(n+1) - 6\Psi(n+\tfrac{1}{2}) + 5\Psi(n+\tfrac{2}{5})\right]$$

$$- \frac{1}{3}\left[\Psi(1) - 6\Psi(\tfrac{1}{2}) + 5\Psi(\tfrac{2}{5})\right].$$

Now it is easily verified from the asymptotic formula that if

$$A + B + C = 0$$

then

$$\lim_{s\to\infty}\left[A\,\Psi(s+a) + B\,\Psi(s+b) + C\,\Psi(s+c)\right] = 0.$$

Hence the sum of the infinite series is given by

$$\sum_{k=1}^{\infty} \frac{1}{(k+1)(2k+1)(5k+2)} = -\frac{1}{3}\left[\Psi(1) - 6\Psi(\tfrac{1}{2}) + 5\Psi(\tfrac{2}{5})\right]$$

$$= 0.03436.$$

A repeated linear factor in the denominator, say $(s+a)^3$, gives rise to terms of the form

$$\frac{A''}{(s+a)^3} + \frac{A'}{(s+a)^2} + \frac{A}{s+a}$$

in the partial fraction representation of $f(x)$. Corresponding to such a factor the solution of the difference equation contains the terms

$$\tfrac{1}{2}A''\Psi''(s+a-1) - A'\Psi'(s+a-1) + A\Psi(s+a-1).$$

An irreducible quadratic factor gives rise to two terms of the form

$$\frac{\alpha + i\beta}{s + a + ib} + \frac{\alpha - i\beta}{s + a - ib}, \qquad i = \sqrt{-1}$$

and the corresponding terms in the solution are

$$\alpha\left[\Psi(s+a-1+ib) + \Psi(s+a-1-ib)\right] + i\beta\left[\Psi(s+a-1+ib) - \Psi(s+a-1-ib)\right].$$

This result, however, is of no great practical value without tables for $\Psi(x)$ with complex arguments.

Case III. Exponential. The solution of

$$\Delta u_s = a^s, \qquad a \neq 1$$

is

$$u_s = \frac{a^s}{a - 1} + w(s)$$

by difference formula 8. Cases involving the product of an exponential and a polynomial may be solved by undetermined coefficients.

Example 4. Solve

$$\Delta u_s = (s^2 - 2s)2^s.$$

Assume

$$u_s = (As^2 + Bs + C)2^s.$$

Then

$$\Delta u_s = \left[As^2 + (4A + B)s + (2A + 2B + C)\right] 2^s.$$

Comparing coefficients we have

$$A = 1; \quad B = -6; \quad C = 10$$

whence

$$u_s = (s^2 - 6s + 10)2^s + w(s).$$

Case IV. Trigonometric, Logarithmic, etc. Difference equations of the type

$$\Delta u_s = \cos ms \quad \text{or} \quad \Delta u_s = \sin ms, \qquad 0 < m < 2\pi$$

are solved by means of difference formulas 11 and 12. The case of polynomial times sine or cosine can be treated by undetermined coefficients in a manner analogous to that of example 4. The details are left to the reader. Simple expressions involving logarithms may be treated by means of difference formula 18.

Exercises

Solve:

1. $\Delta u_s = s^2 - 3s + 2.$

2. $\Delta u_s = \frac{2}{s(s + 1)}.$

3. $\Delta u_s = sx^s.$

4. $\Delta u_s = \sin \frac{\pi s}{6}.$

5. $\Delta u_s = \frac{1}{(2s + 1)(5s + 3)}.$

Find the sum of n terms of the series:

6. $1\cdot 2\cdot 3 + 2\cdot 3\cdot 4 + 3\cdot 4\cdot 5 + 4\cdot 5\cdot 6 + \ldots .$

7. $\frac{2}{1\cdot 3} + \frac{3}{2\cdot 4} + \frac{4}{3\cdot 5} + \frac{5}{4\cdot 6} \ldots .$

8. $x + 2x^2 + 3x^3 + 4x^4 + 5x^5 + \ldots .$

Find the sum of the infinite series:

9. $\frac{1}{1\cdot 2\cdot 3} + \frac{1}{2\cdot 3\cdot 4} + \frac{1}{3\cdot 4\cdot 5} + \frac{1}{4\cdot 5\cdot 6} \ldots .$

10. $\frac{1\cdot 2}{4} + \frac{2\cdot 3}{4^2} + \frac{3\cdot 4}{4^3} + \frac{4\cdot 5}{4^4} + \ldots .$

11. $\frac{1}{1\cdot 3\cdot 9} + \frac{3}{2\cdot 5\cdot 14} + \frac{5}{3\cdot 7\cdot 19} + \frac{7}{4\cdot 9\cdot 24} + \ldots .$

12. Solve $\Delta u_s = \ln \left[a(s + b)^n\right].$

13. Find the sum $\sum_{n=1}^{100} \ln n.$

86. EXACT EQUATIONS

The method of separation of variables, so important for differential equations, is of little value for difference equations. The reason is that in our difference formulas the Δs is constant (and in fact was chosen to be unity). But the Δu is not constant and we have no way of summing expressions containing Δu. Thus the equation

$$\Delta u = a^s u$$

could be put in the form

$$\sum \frac{\Delta u}{u} = \sum a^s$$

but since we do not know how to evaluate $\sum \frac{\Delta u}{u}$ for variable Δu, we are unable to complete the solution.

If, however, it is possible to express a difference equation in the form of a difference equated to zero, the solution can be obtained at once.

Example 1. The difference equation

$$(u_{s+1} - u_s)(u_{s+1} + u_s + 2s) + 2(u_s - s) = 0$$

may be written as a difference

$$\left[u_{s+1}^2 + 2su_{s+1} - s(s+1)\right] - \left[u_s^2 + 2(s-1)u_s - s(s-1)\right] = 0$$

which is in the form

$$F(u_{s+1}, s+1) - F(u_s, s) = 0.$$

The solution is therefore

$$F(u_s, s) = w(s)$$

or in this case

$$u_s^2 + 2(s-1)u_s - s(s-1) = w(s).$$

Example 2. The linear difference equation

$$u_{s+1} - au_s = 0, \qquad a \neq 1,$$

is not exact as it stands. However, multiplication by the factor a^{-s-1} puts it in the exact form

$$a^{-(s+1)} u_{s+1} - a^{-s} u_s = 0$$

which has the solution

$$a^{-s} u_s = w(s)$$

or

$$u_s = a^s w(s).$$

Example 3. The linear difference equation

$$u_{s+1} - \frac{(s+a)(s+b)}{(s+c)} u_s = 0$$

may be made exact by the factor

$$\frac{\Gamma(s+1+c)}{\Gamma(s+1+a) \quad \Gamma(s+1+b)}.$$

For by relations $\Gamma(s+1+c) = (s+c) \quad \Gamma(s+c)$, etc., the resulting equation is

$$\frac{\Gamma(s+1+c)}{\Gamma(s+1+a) \quad \Gamma(s+1+b)} u_{s+1} - \frac{\Gamma(s+c)}{\Gamma(s+a) \quad \Gamma(s+b)} u_s = 0$$

and the solution is

$$u_s = \frac{\Gamma(s+a) \quad \Gamma(s+b)}{\Gamma(s+c)} w(s).$$

The generalization of this example is obvious.

Example 4. The non-homogeneous linear difference equation

$$u_{s+1} - au_s = bs$$

can be made exact by the factor $a^{-(s+1)}$. For we have

$$a^{-(s+1)} u_{s+1} - a^{-s} u_s = bsa^{-s-1}$$

where the left member is already exact and the right-hand member can be made exact by undetermined coefficients as in Example 4, Article 85. We then have

$$a^{-(s+1)} u_{s+1} - a^{-s} u_s = -b\left[\frac{s+1}{a-1} + \frac{1}{(a-1)^2}\right]a^{-s-1} + b\left[\frac{s}{a-1} + \frac{1}{(a-1)^2}\right]a^{-s}.$$

This gives

$$a^{-s} u_s = -b\left[\frac{s}{a-1} + \frac{1}{(a-1)^2}\right]a^{-s} + w(s)$$

whence

$$u_s = a^s w(s) - \frac{bs}{a-1} - \frac{b}{(a-1)^2}.$$

Exercises

Obtain the general solution of the following difference equations.

1. $u_{s+1} - su_s = 0.$

2. $3u_{s+1} - u_s = s.$

3. $u_{s+1}^2 - su_s^2 = 0.$

4. $su_{s+1} - au_s = 0.$

5. $\ln\left(\frac{3u_{s+1}}{su_s}\right) = 2.$

6. $u_{s+1} - au_s = b^s.$

7. $su_{s+1} - (s+1)u_s = 1.$

8. $u_{s+1} - (as + b)u_s = 0.$

9. $u_{s+1} = asu_s.$

10. $u_{s+1} = \frac{as+b}{cs+d}\, u_s.$

11. $u_{s+1} = s^{(n)}\, u_s.$

12. $u_{s+1} + su_s = 0.$

87. LINEAR DIFFERENCE EQUATIONS OF ORDER HIGHER THAN THE FIRST

Among difference equations of order two or more, the only type which we can treat with any degree of completeness is the homogeneous linear difference equation with constant coefficients. The procedure will be illustrated by means of examples.

Example 1. Consider the difference equation

$$u_{s+2} - 7u_{s+1} + 10u_s = 0.$$

In analogy with homogeneous linear differential equations with constant coefficients we assume a solution

$$u_s = a^s$$

where a is to be determined. Substitution in the differential equation and removal of the factor a^s gives the equation

$$a^2 - 7a + 10 = 0$$

whence $a = 2$, or $a = 5$. Two particular solutions are therefore

$$u_s = 2^s \quad \text{and} \quad u_s = 5^s.$$

The reader may easily verify that <u>if y_s and z_s are two solutions of any homogeneous linear difference equation then</u>

$$u_s = y_s w_1 + z_s w_2,$$

<u>where w_1 and w_2 are arbitrary periodic functions of period 1, is also a solution.</u> In the present example

$$u_s = w_1 2^s + w_2 5^s$$

is a solution and in fact the general solution.

<u>Example 2</u>. A similar treatment of the difference equation

$$u_{s+3} - 6u_{s+2} + 12u_{s+1} - 8u_s = 0$$

leads to the equation

$$a^3 - 6a^2 + 12a - 8 = 0,$$

and this has three roots each equal to 2. One particular solution is therefore

$$u_s = 2^s.$$

By analogy with the similar case in differential equations we expect to find that $u_s = s2^s$, and $u_s = s^{(2)}2^s$ are also solutions, and this expectation is proved by actual substitution to be justified. The general solution is therefore

$$u_s = \left[w_1 + sw_2 + s^{(2)}w_3\right] 2^s.$$

Example 3. The substitution $u_s = a^s$ in the difference equation

$$u_{s+2} - 4u_{s+1} + 13u_s = 0$$

gives

$$a^2 - 4a + 13 = 0,$$

with roots

$$a = 2 \pm 3i.$$

Hence the general solution is

$$u_s = w_1(2 + 3i)^s + w_2(2 - 3i)^s.$$

To avoid complex numbers we may express the solution in trigonometric form as

$$u_s = 13^{\frac{s}{2}} \left[w_3 \cos s\theta + w_4 \sin s\theta\right],$$

where θ is the principal value $(-\frac{\pi}{2} < \theta \leqq \frac{\pi}{2})$ of either arc tan $(\frac{3}{2})$ or arc tan $(-\frac{3}{2})$.

A few cases of nonhomogeneous linear difference equations with constant coefficients may be solved by means of undetermined coefficients. Just as in the case of linear differential equations the general solution of the nonhomogeneous difference equation is found by adding a particular solution of the nonhomogeneous equation to the general solution of the associated homogeneous equation.

Example 4. To solve the equation

$$u_{s+2} - 7u_{s+1} + 10u_s = 12 \cdot 4^s,$$

we first obtain the general solution of

$$u_{s+2} - 7u_{s+1} + 10u_s = 0.$$

This is the equation of Example 1, and the general solution is

$$u_s = w_1 2^s + w_2 5^s.$$

Next we look for a particular solution of the original nonhomogeneous equation and to this end assume

$$u_s = A4^s$$

in which A is a constant to be determined. Substitution in the difference equation gives

$$A4^s \left[4^2 - 7 \cdot 4 + 10\right] = 12 \cdot 4^s$$

whence $A = -6$. The general solution is

$$u_s = w_1 2^s + w_2 5^s - 6 \cdot 4^s.$$

<u>Example 5</u>. The method of Example 4 fails if the right-hand member satisfies the homogeneous equation, as in the equation

$$u_{s+2} - 7u_{s+1} + 10u_s = 12 \cdot 5^s.$$

In such a case we assume a particular solution of the form

$$u_s = As \cdot 5^s$$

and find that $A = \frac{4}{5}$. The solution is

$$u_s = w_1 2^s + w_2 5^s + \tfrac{4}{5} s \cdot 5^s.$$

Example 6. The procedure for the case where the right-hand member is a polynomial may be illustrated by the example

$$u_{s+2} - 4u_s = 9s^2.$$

The solution of the homogeneous equation is

$$u_s = w_1 2^s + w_2(-2)^s.$$

For the particular solution we assume

$$u_s = As^2 + Bs + C$$

and find that the equation is satisfied if $A = -3$, $B = -4$, $C = -20/3$, and the general solution is

$$u_s = w_1 2^s + w_2(-2)^s - 3s^2 - 4s - 20/3.$$

The foregoing examples should suffice to guide any reader familiar with linear differential equations to an understanding of linear difference equations with constant coefficients, and to suggest methods of attacking cases not illustrated here.

Exercises

Solve:

1. $\Delta^2 u_{s-1} = u_s$

2. $\Delta^2 u_{s-1} + u_s = 0$

3. $u_{s+2} + 2u_{s+1} + 2u_s = 0$

4. $u_s = pu_{s+1} + qu_{s-1}, \quad (p + q = 1).$

5. $u_{s+2} - 4u_{s+1} + 4u_s = 0$

6. $u_{s+4} - 2u_{s+2} + u_s = 0$

7. $u_{s+2} + a^2u_s = a^s$

8. $u_{s+2} - a^2u_s = a^s$

9. $u_{s+2} - 4u_s = s.$

10. $u_{s+2} - u_s = s.$

11. $u_{s+2} - 2(s + 1)u_{s+1} + s(s + 1)u_s = 0$

(Hint: Let $u_s = v_s \Gamma(s)$).

12. $s(s + 1)u_{s+2} - 2su_{s+1} + u_s = 0.$

88. LINEAR EQUATIONS, VARIABLE COEFFICIENTS

A solution of a linear difference equation with variable coefficients can frequently be obtained by means of factorial series of the form

$$u_s = a_0 + a_1 s + a_2 s^{(2)} + a_3 s^{(3)} + a_4 s^{(4)} + \ldots$$

When s is a positive integer such a series always terminates, hence if the use of these series is restricted to positive integral values of s, the troublesome question of convergence does not arise.

Example. Obtain a solution of the difference equation

$$(n - s)u_{s+1} + (2t - n)u_s + su_{s-1} = 0$$

in which n and t are positive integers, $t \leq n$.

Assume

$$u_s = \sum_{k=0}^{\infty} a_k s^{(k)}.$$

Then

$$su_{s-1} = \sum a_k s^{(k+1)}$$

$$u_{s+1} = \sum a_k s^{(k)} + \sum k a_k s^{(k-1)}$$

$$su_{s+1} = \sum a_k s^{(k+1)} + 2\sum k a_k s^{(k)} + \sum k(k-1)a_k s^{(k-1)}.$$

When these expressions are substituted into the difference equation and the coefficients of factorials of like degree are collected, it turns out that the coefficient of $s^{(k)}$ is

$$(k+1)(n-k)a_{k+1} + 2(t-k)a_k.$$

Since these coefficients must all vanish we have

$$a_{k+1} = -\frac{2(t-k)}{(k+1)(n-k)} a_k.$$

A solution of this first order difference equation is evidently

$$a_k = (-1)^k \frac{2^k t^{(k)}}{k!\, n^{(k)}} a_o.$$

When these coefficients are inserted in the series assumed for u_s, the result is

$$u_s = a_o \sum (-1)^k \frac{2^k t^{(k)} s^{(k)}}{k!\, n^{(k)}}$$

Since t was assumed to be an integer not greater than n it is evident that the series terminates with the term in which $k = t$, and that u_s is simply a polynomial in s of degree t. In fact a comparison of this solution with equation (2), Article 82, shows that

$$u_s = a_o \frac{H_t(s)}{\binom{n}{s}}$$

It should be noted that only one solution was obtained in this instance although the difference equation actually has two independent particular solutions. A critical analysis of this and other peculiarities that may be encountered is beyond the scope of this elementary treatment. The reader will find the subject adequately discussed in Milne-Thompson, The Calculus of Finite Differences, Chapter XIV.

Exercises

Find at least one solution of each of the following difference equations.

(1) $u_{x+2} - (x + 1)u_{x+1} + (x + 6)u_x = 0$

(2) $u_{x+2} + (x + 2)(x + 1)u_x = 0$

(3) $(x + 2)u_{x+2} - (2x + 3)u_{x+1} + (2x + 2)u_x = 0$

(4) $(x + 2)u_{x+2} - xu_{x+1} - u_x = 0$

APPENDICES

APPENDIX A
NOTATION AND SYMBOLS

In the hope of minimizing one of the sources of confusion that plague beginners we have kept the number of new symbols in the text as small as possible. Since, however, many other symbols are current in the literature, reference is here made to a few of the more important ones.

1. The symbol Δ defined by $\Delta u_x = u_{x+1} - u_x$ is fairly standard.
   However, some writers restrict Δ to the case where the interval between values of the independent variable is unity and use $\underset{h}{\Delta}$ for the general case where the interval is h.
2. The symbol ∇ defined by $\nabla u_x = u_x - u_{x-1}$ is occasionally used for backward differences.
3. The displacement operator E defined by $E = 1 + \Delta$ or $Eu_x = u_{x+1}$ is much used by writers on finite differences, and is really very handy.
4. Sheppard's central difference notation defined by $\delta u_x = u_{x+1/2} - u_{x-1/2}$ and $\mu u_x = \frac{1}{2}(u_{x+1/2} + u_{x-1/2})$ is particularly useful in the symbolic treatment of central difference formulas.
5. <u>Notation for divided differences</u>. The divided differences belonging to the arguments x_0, x_1, x_2, x_3, . . ., and functional values $f(x_0)$, $f(x_1)$, $f(x_2)$, . . . or u_0, u_1, u_2, . . . have been represented by the various notations
   (1) $[x_0, x_1]$, $[x_0, x_1, x_2]$, etc., as in the text and in Milne-Thomson and Fort.
   (2) (x_0, x_1), (x_0, x_1, x_2), etc. Sheppard.
   (3) $f(x_0, x_1)$, $f(x_0, x_1, x_2)$, etc., Steffensen.

(4) $\underset{x_1}{\Delta} u_{x_0}$, $\underset{x_1 x_2}{\Delta^2} u_{x_0}$, etc. Aitken, Freeman.

(5) $\mathcal{D}f(x_0)$, $\mathcal{D}^2 f(x_0)$, etc. Jordan

(6) $\Delta(x_0, x_1)$, $\Delta^2(x_0, x_1, x_2)$, etc., Fraser

(7) θu_0, $\theta^2 u_0$, etc., De Morgan.

It is possible to raise objections to every one of these, and the author presents no argument in defense of his choice.

APPENDIX B
TEXTS, TABLES, AND BIBLIOGRAPHIES

For readers who desire a broader grasp of the subject or information on topics not treated here (osculatory interpolation, interpolation with two or more variables, theory of errors, Graeffe's root squaring method--to mention just a few) may be interested in the following short list of books and pamphlets. The nature of the work is briefly indicated.

TEXTS

Bennett, A. A., Milne, W. E., Bateman, Harry. Numerical Integration of Differential Equations. Report of National Research Council Committee on Numerical Integration. National Academy of Sciences, Washington, D. C., 1933.

The Interpolating Polynomial, Successive Approximation, Step-by-step Methods of Integration, Methods for Partial Differential Equations.

Boole, George, Finite Differences, 2nd Ed. 1872 (3rd Ed. is reprint of 2nd Ed., New G. E. Steckert and Company, 1931).

A classic on finite differences and their applications.

Fort, Tomlinson, Finite Differences and Difference Equations in the Real Domain, Oxford University Press, 1948.

Freeman, Harry, Mathematics for Actuarial Students, vol. II, Cambridge University Press, 1939.

As title indicates, it is primarily for actuarial students but is quite broad in scope.

Glover, J. W., Interpolation, Summation, and Graduation, Chapter 3 of Handbook of Mathematical Statistics, (H. L. Rietz, Ed.-in-chief), pp. 34-61, Boston, Houghton Mifflin, 1924.

A condensed 28-page summary of the features most likely

to prove of use to the student of statistics, having an actuarial bent.

Jordan, Charles, *Calculus of Finite Differences*, Budapest, 1939. Hungarian Agent Eggenberger Book Shop.

For the advanced reader.

Kowalewski, G., *Interpolation und Genäherte Quadratur*, B. G. Teubner, Leipzig und Berlin, 1932.

A compact and well-written book on interpolation and numerical integration for more advanced readers.

Levy, H. and Baggott, E. A., *Numerical Studies in Differential Equations*, Watts and Company, London.

Comprehensive account of nearly all methods available for the numerical solution of differential equations.

Lipka, Joseph, *Graphical and Mechanical Computation*, New York, Wiley, 1918. Chapter 7, Empirical Formulas--periodic curves, pp. 170-208; Chapter 8, Interpolation, pp. 209-223; Chapter 9, Approximate Integration and Differentiation, pp. 224-259.

A brief discussion of some standard methods with illustrative examples and collections of exercises for practice.

Markoff, A. A., *Differenzenrechnung*. German Translation. Leipzig, Teubner, 1896. Chapter 3, Relation Between Differences and Differential Quotients, pp. 20-27; Chapter 4, Construction and Use of Mathematical Tables, pp.28-49; Chapter 5, Application of Interpolation to the Computation of Integrals, pp. 50-71; Chapter 9, Euler's formula, pp.112-125; Chapter 10, Applications of Euler's Formula, pp. 126-140.

Some interesting numerical examples, and some original points of view. Sketchy as to theory of interpolation.

Miller, Morton D., et al., *Elements of Graduation*. Published by the Actuarial Society of America and the American Institute of Actuaries.

Milne-Thompson, L. M., *Calculus of Finite Differences*, Macmillan and Company, Ltd., London, 1933.

Excellent treatment of differences, summation, and difference equations. Primarily for advanced readers.

Nörlund, N. E., *Differenzenrechnung*. Berlin, Springer, 1924 Chapter 1, Fundamental Notions, pp. 3-16; Chapter 2, The Bernoulli and Euler Polynomials, pp. 17-37; Chapter 7, Multiple Sums, pp. 163-197; Chapter 8, Interpolational Series, pp. 198-256; Chapter 15, Reciprocal Differences and Continued Fractions, pp. 415-45

The standard treatise on newer aspects of difference equations. Classical interpolation, while incidental to the subject, is handled excellently.

Pearson, Karl, Tracts for Computers, No. 2. On the Construction of Tables and on Interpolation. Part 1. Uni-variate Tables. Cambridge, Univ. Press, 1920. (Multivariate Tables are considered in Tract No. 3.)

A forceful, original, and direct discussion of some theoretical and practical problems facing the table-computer. Appendix II, some works and memoirs dealing with interpolation, gives 50 titles chiefly of historical interest, with evaluating remarks on each.

Runge, Carl, and König, H., Vorlesungen über numerisches Rechnen. Berlin, Springer, 1924: Chapter 4, Entire Rational Functions, in particular pp. 103-114; Chapter 9, Numberical Integration and Differentiation, pp. 238-285; Chapter 10, Numerical Integration of Ordinary Differential Equations, pp. 286-325.

A much-quoted standard treatise.

Runge, Carl, and Willers, F. A., Numerische und graphische Quadratur und Integration gewöhnlicher und partieller Differentialgleichungen. In Encyclopädie der mathematischen Wissenschaften, Band 2 (3), Heft 2., pp. 47-176. Leipzig, Teubner, 1915.

Scarborough, James B., Numerical Mathematical Analysis, Baltimore, Johns Hopkins Press, 1930.

Excellent for beginners.

Sheppard, W. F., Interpolation. Encyclopedia Britannica, 14th ed., 1929, vol. 12, pp. 354-357; Mensuration, ibid., vol. 15, pp. 253-257.

Excellent, readable, concise, summaries of first principles.

Steffensen, J. F., Interpolation. Baltimore, Williams and Wilkins, 1927.

A standard treatise on the classical parts of the theory. Original in method. Comprehensive.

von Sanden, H., Practical Mathematical Analysis, (Translated by H. Levy.), New York, E. P. Dutton and Company, 1913.

Graphical and Numerical methods presented on an elementary level.

Whittaker, E. T. and Robinson, G., The Calculus of Observations, Blackie and Son, Ltd., London and Glascow, 1932.

Contains a wealth of valuable material. One of the best books for the general reader.

TABLES

A short list of mathematical tables likely to be useful to computers is supplied for reference.

Anderson, R. L. and Houseman, E. E., *Tables of Orthogonal Polynomial Values Extended to N = 104*. Agricult. Exper. Station, Iowa State Coll. of Agricult. Mech. Arts, Statist. Sect., Res. Bull. No. 297, pp. 595-672 (1942).

Barlow's Tables of Squares, Cubes, etc., E. and F. N. Spon, Ltd. London, 1930.

A useful general purpose table for any computer.

Boll, Marcel, *Tables Numériques Universelles des Laboratoires et Bureaux d'Etude*. Dunod, Paris, 1947, ii + 882. pp.

Blaisher, J. W. L., Bickley, W. G., Gwyther, C. E., Miller, J. C. P., and Ternouth, E. J., *Table of Powers Giving Integral Powers of Integers*, British Association for the Advancement of Science, Mathematical Tables, vol. 9. Cambridge University Press, Cambridge, England; Macmillan Company, New York, 1940, xii + 132. pp.

Glover, James W., *Tables of Applied Mathematics in Finance, Insurance, Statistics*, George Wahr, Pub., Ann Arbor, 1930.

Contains in addition to tables of finance, etc., five-place tables of the functions $\varphi_0, \varphi_1, \ldots, \varphi_8$ of Art. 81, five-place Tables of Newton's Bessel's, and Everett's Interpolation Coefficients to fifth differences. Five-place Tables of the derivatives of Newton's Bessel's, and Everett's interpolation coefficients. Many other special tables of use to computers.

Mathematical Tables, Part-volume A. Legendre Polynomials. Cambridge, University Press, New York, The Macmillan Company, 1946, pp.1-42.

Mathematical Tables, Vol. 1. British Ass'n. for the Advancement of Science, Cambridge University Press.

Contains among other tables the gamma, digamma, trigamma, tetragamma, and pentagamma functions to ten places. Also tables of $Hh_{-n}(x) = (-1)^n \sqrt{2\pi}\,\varphi_{n-1}(x)$ with accuracy varying from six to ten places.

Table of the First Ten Powers of the Integers from 1 to 1000. Published under the sponsorship of the

National Bureau of Standards as a report of Official Project No. 365-97-3-11, conducted under the auspices of the United States Works Progress Administration for the City of New York, 1939, viii + 80 pp.

Table of Reciprocals of the Integers from 100,000 through 200,009. Math. Tables Project.

Tables of Fractional Powers. Math. Tables Project. Columbia University Press, New York, 1946.

Tables of Lagrangian Interpolation Coefficients. Columbia University Press, New York, 1944.

Tables of Probability Functions. Math. Tables Project. Values of $\varphi_o(x)$ (See Art. 81) and $\int_{-x}^{x} \varphi_o(x)\, dx$ to fifteen decimal places at intervals of 0.0001 for $x < 1$ and 0.001 for $x > 1$.

Thompson, A. J., _Table of the Coefficients of Everett's Central-Difference Interpolation Formula_. Tracts for Computers, No. 5, 2nd edition. Cambridge University Press, Cambridge, England, 1943. viii + 32 pp.

TABLES IN JOURNALS

Huntington, E. V., Tables of Lagrangian coefficients for Interpolating Without Differences. _Proc. Am. Acad. Arts_ Sci. 63, pp. 421-437, 1929.

Lowan, A. N., and Salzer, Herbert., Table of coefficients in Numerical Integration Formulae. _J. Math. Phys._, Mass. Inst. Tech. 22, pp. 49-50, 1943.

Lowan, Arnold N., Davids, Norman, and Levenson, Arthur, Table of the Zeros of the Legendre Polynomials of Order 1-16 and the Weight Coefficients for Gauss's Mechanical Quadrature Formula. _Bull. Amer. Soc._ 48, pp. 739-743, 1942.

Lowan, Arnold N., and Salzer, Herbert E., Table of Coefficients for Numerical Integration Without Differences. _J. Math. Phys._, Mass. Inst. Tech. 24, 1-21, 1945.

Lowan, Arnold N., Salzer, Herbert E., and Hillman, Abraham. A Table of Coefficients for Numerical Differentiation. _Bull. Amer. Math. Soc._ 48, pp. 920-924, 1942.

Rutledge, G., and Crout, P. D., Tables and Methods for Extending Tables for Interpolation Without Differences. Pub. Mass. Inst. of Tech. (Dept. of Math.) Ser. 2, No. 178, Sept., 1930.

Salzer, Herbert E., Coefficients for Numerical Differentiation with Central Differences. J. Math. Phys., Mass. Inst. Tech. 22, pp. 115-135, 1943.

Salzer, Herbert E., Table of Coefficients for Double Quadrature Without Differences, for Integrating Second Order Differential Equations. J. Math. Phys., Mass. Inst. Tech. 24, pp. 135-140, 1945.

Salzer, Herbert E., Table of Coefficients for Inverse Interpolation With Central Differences. J. Math. Phys., Mass. Inst. Tech. 22, pp. 210-224, 1943.

BIBLIOGRAPHIES

For more complete coverage of the literature the reader may consult the bibliographies cited below.

Adams, C. R., Bibliography, supplementary to Nörlund's bibliography on the Calculus of Finite Differences and Difference Equations. Bull. Am. Math. Soc. (2) 37: pp. 383-400, 1931.

Brings Nörlund's bibliography up to date (1930).

Bennett, A. A., Milne, W. E., Bateman, Harry. Numerical Integration of Differential Equations, National Research Council Committee on Numerical Integration. The National Academy of Sciences, Washington, D. C.

Contains a carefully chosen and fairly extensive bibliography on numerical methods.

Fletcher, A., Miller, J. C. P., and Rosenhead, L. An Index of Mathematical Tables. McGraw-Hill Book Company, New York; Scientific Computing Service Limited, London, 1946, viii + 451 pp.

A monumental work in two parts. Part I, 350 pages, gives a classified list of tables. Part II, 72 pages, gives a bibliography.

Nörlund, N. E. Differenzenrechnung. Berlin, Springer, 1924.

Contains bibliography with 500 authors and 1373 works cited.

APPENDIX C

CLASSIFIED GUIDE TO FORMULAS AND METHODS

The occasional reader who desires a formula or method suited to a particular problem without searching through the entire text may be aided by the following guide. Topics which are adequately described by chapter titles (e.g., solution of differential equations, difference equations, etc.) have not been included here.

Interpolation

I. Polynomial Interpolation	Articles
1. Using repeated linear interpolation with either equally or unequally spaced points	20, 21
2. Using ordinates at unequally spaced points	24
3. Using ordinates at equally spaced points	25
4. Using differences with equally spaced points	
a) Near beginning of table	47 (3)
b) Near end of table	48 (3)
c) Elsewhere in table	49 (2), (3) 50 (2), 52
d) For tables providing even ordered central differences	51
5. Using divided differences with unequally spaced points	Chapter VIII, especially Art. 60, 61

II. Interpolation by Rational Fractions
(Used when given function has poles near required point.)

For equally or unequally spaced points — Chapter VIII, especially Art. 64

III. Trigonometric Interpolation
(For periodic functions.)

1. For brief mention see 18

2. For general case, equally spaced points 79, 80

(Note that when $n = k$ the Least Square approximation fits exactly at the given points and becomes an interpolating formula.)

Numerical Differentiation

Numerical Integration

the most accurate formulas available.)

6. Using <u>differences</u> with equally spaced points — 56, 57, 58

Least Square Approximations

	Articles
I. <u>Minimizing Sum of Squares of Errors at Discrete Points</u>	
1. <u>Polynomial</u> Approximation	
a) Unequally spaced points,	67, 68
b) Equally spaced points	72, 73
2. <u>Trigonometric</u> Approximation	
Equally spaced points	79, 80
3. <u>Analog</u> of Gram-Charlier Approximation	
Equally spaced points	82
II. <u>Minimizing Integral of Square of Error</u>	
1. <u>Polynomial</u> Approximation	69, 70, 71
2. <u>Trigonometric</u> Approximation	78
3. <u>Gram-Charlier</u> Approximation	81
4. General weighted Approximations	76

Summation

For elementary examples of summation see Articles 43 and 85.

TABLES

Table I

BINOMIAL COEFFICIENTS $\binom{n}{k}$

n \ k	0	1	2	3	4	5	6	7	8	9	10
1	1	1									
2	1	2	1								
3	1	3	3	1							
4	1	4	6	4	1						
5	1	5	10	10	5	1					
6	1	6	15	20	15	6	1				
7	1	7	21	35	35	21	7	1			
8	1	8	28	56	70	56	28	8	1		
9	1	9	36	84	126	126	84	36	9	1	
10	1	10	45	120	210	252	210	120	45	10	1
11	1	11	55	165	330	462	462	330	165	55	11
12	1	12	66	220	495	792	924	792	495	220	66
13	1	13	78	286	715	1287	1716	1716	1287	715	286
14	1	14	91	364	1001	2002	3003	3432	3003	2002	1001
15	1	15	105	455	1365	3003	5005	6435	6435	5005	3003
16	1	16	120	560	1820	4368	8008	11440	12870	11440	8008
17	1	17	136	680	2380	6188	12376	19448	24310	24310	19448
18	1	18	153	816	3060	8568	18564	31824	43758	48620	43758
19	1	19	171	969	3876	11628	27132	50388	75582	92378	92378
20	1	20	190	1140	4845	15504	38760	77520	125970	167960	184756

Table II

INTERPOLATION COEFFICIENTS $\binom{s}{k}$ FOR

NEWTON'S BINOMIAL INTERPOLATION FORMULA

s	$\binom{s}{2}$	$\binom{s}{3}$	$\binom{s}{4}$	$\binom{s}{5}$	s
.00	.00000	.00000	.00000	.00000	.00
.01	-.00495	.00328	-.00245	.00196	.01
.02	-.00980	.00647	-.00482	.00384	.02
.03	-.01455	.00955	-.00709	.00563	.03
.04	-.01920	.01254	-.00928	.00735	.04
.05	-.02375	.01544	-.01139	.00899	.05
.06	-.02820	.01824	-.01340	.01056	.06
.07	-.03255	.02094	-.01534	.01206	.07
.08	-.03680	.02355	-.01719	.01348	.08
.09	-.04095	.02607	-.01897	.01483	.09
.10	-.04500	.02850	-.02066	.01612	.10
.11	-.04895	.03084	-.02228	.01733	.11
.12	-.05280	.03309	-.02382	.01849	.12
.13	-.05655	.03525	-.02529	.01958	.13
.14	-.06020	.03732	-.02669	.02060	.14
.15	-.06375	.03931	-.02801	.02157	.15
.16	-.06720	.04122	-.02926	.02247	.16
.17	-.07055	.04304	-.03045	.02332	.17
.18	-.07380	.04477	-.03156	.02412	.18
.19	-.07695	.04643	-.03261	.02485	.19
.20	-.08000	.04800	-.03360	.02554	.20
.21	-.08295	.04949	-.03452	.02617	.21
.22	-.08580	.05091	-.03538	.02675	.22
.23	-.08855	.05224	-.03618	.02728	.23
.24	-.09120	.05350	-.03692	.02776	.24
.25	-.09375	.05469	-.03760	.02820	.25
.26	-.09620	.05580	-.03822	.02859	.26
.27	-.09855	.05683	-.03879	.02893	.27
.28	-.10080	.05779	-.03930	.02924	.28
.29	-.10295	.05868	-.03976	.02950	.29
.30	-.10500	.05950	-.04016	.02972	.30
.31	-.10695	.06025	-.04052	.02990	.31
.32	-.10880	.06093	-.04082	.03004	.32
.33	-.11055	.06154	-.04108	.03015	.33
.34	-.11220	.06208	-.04129	.03022	.34
.35	-.11375	.06256	-.04145	.03026	.35

s	$\binom{s}{2}$	$\binom{s}{3}$	$\binom{s}{4}$	$\binom{s}{5}$	s
.35	-.11375	.06256	-.04145	.03026	.35
.36	-.11520	.06298	-.04156	.03026	.36
.37	-.11655	.06333	-.04164	.03023	.37
.38	-.11780	.06361	-.04167	.03017	.38
.39	-.11895	.06384	-.04165	.03007	.39
.40	-.12000	.06400	-.04160	.02995	.40
.41	-.12095	.06410	-.04151	.02980	.41
.42	-.12180	.06415	-.04138	.02962	.42
.43	-.12255	.06413	-.04121	.02942	.43
.44	-.12320	.06406	-.04100	.02919	.44
.45	-.12375	.06394	-.04076	.02894	.45
.46	-.12420	.06376	-.04049	.02866	.46
.47	-.12455	.06352	-.04018	.02836	.47
.48	-.12480	.06323	-.03984	.02804	.48
.49	-.12495	.06289	-.03946	.02770	.49
.50	-.12500	.06250	-.03906	.02734	.50
.51	-.12495	.06206	-.03863	.02696	.51
.52	-.12480	.06157	-.03817	.02657	.52
.53	-.12455	.06103	-.03769	.02615	.53
.54	-.12420	.06044	-.03717	.02572	.54
.55	-.12375	.05981	-.03664	.02528	.55
.56	-.12320	.05914	-.03607	.02482	.56
.57	-.12255	.05842	-.03549	.02434	.57
.58	-.12180	.05765	-.03488	.02386	.58
.59	-.12095	.05685	-.03425	.02336	.59
.60	-.12000	.05600	-.03360	.02285	.60
.61	-.11895	.05511	-.03293	.02233	.61
.62	-.11780	.05419	-.03224	.02180	.62
.63	-.11655	.05322	-.03154	.02125	.63
.64	-.11520	.05222	-.03081	.02071	.64
.65	-.11375	.05119	-.03007	.02015	.65
.66	-.11220	.05012	-.02932	.01958	.66
.67	-.11055	.04901	-.02855	.01901	.67
.68	-.10880	.04787	-.02777	.01844	.68
.69	-.10695	.04670	-.02697	.01785	.69
.70	-.10500	.04550	-.02616	.01727	.70

II. INTERPOLATION COEFFICIENTS $\binom{s}{k}$ FOR NEWTON'S BINOMIAL INTERPOLATION FORMULA

s	$\binom{s}{2}$	$\binom{s}{3}$	$\binom{s}{4}$	$\binom{s}{5}$	s
.70	-.10500	.04550	-.02616	.01727	.70
.71	-.10295	.04427	-.02534	.01668	.71
.72	-.10080	.04301	-.02451	.01608	.72
.73	-.09855	.04172	-.02368	.01548	.73
.74	-.09620	.04040	-.02283	.01488	.74
.75	-.09375	.03906	-.02197	.01428	.75
.76	-.09120	.03770	-.02111	.01368	.76
.77	-.08855	.03631	-.02024	.01308	.77
.78	-.08580	.03489	-.01937	.01247	.78
.79	-.08295	.03346	-.01848	.01187	.79
.80	-.08000	.03200	-.01760	.01126	.80
.81	-.07695	.03052	-.01671	.01066	.81
.82	-.07380	.02903	-.01582	.01006	.82
.83	-.07055	.02751	-.01493	.00946	.83
.84	-.06720	.02598	-.01403	.00887	.84
.85	-.06375	.02444	-.01314	.00828	.85
.86	-.06020	.02288	-.01224	.00769	.86
.87	-.05655	.02130	-.01134	.00710	.87
.88	-.05280	.01971	-.01045	.00652	.88
.89	-.04895	.01811	-.00955	.00594	.89
.90	-.04500	.01650	-.00866	.00537	.90
.91	-.04095	.01488	-.00777	.00480	.91
.92	-.03680	.01325	-.00689	.00424	.92
.93	-.03255	.01161	-.00601	.00369	.93
.94	-.02820	.00996	-.00513	.00314	.94
.95	-.02375	.00831	-.00426	.00260	.95
.96	-.01920	.00666	-.00339	.00206	.96
.97	-.01455	.00500	-.00254	.00154	.97
.98	-.00980	.00333	-.00168	.00102	.98
.99	-.00495	.00167	-.00084	.00050	.99
1.00	.00000	.00000	.00000	.00000	1.00

Table III

EVERETT'S INTERPOLATION COEFFICIENTS

s	$\binom{s+1}{3}$	$\binom{s+2}{5}$	s	$\binom{s+1}{3}$	$\binom{s+2}{5}$
.00	.00000	.00000	.25	-.03906	.00769
.01	-.00167	.00033	.26	-.04040	.00794
.02	-.00333	.00067	.27	-.04172	.00819
.03	-.00500	.00100	.28	-.04301	.00843
.04	-.00666	.00133	.29	-.04427	.00867
.05	-.00831	.00166	.30	-.04550	.00890
.06	-.00996	.00199	.31	-.04670	.00912
.07	-.01161	.00232	.32	-.04787	.00933
.08	-.01325	.00265	.33	-.04901	.00954
.09	-.01488	.00297	.34	-.05012	.00973
.10	-.01650	.00329	.35	-.05119	.00992
.11	-.01811	.00361	.36	-.05222	.01011
.12	-.01971	.00393	.37	-.05322	.01028
.13	-.02130	.00424	.38	-.05419	.01045
.14	-.02288	.00455	.39	-.05511	.01060
.15	-.02444	.00486	.40	-.05600	.01075
.16	-.02598	.00516	.41	-.05685	.01089
.17	-.02751	.00546	.42	-.05765	.01102
.18	-.02903	.00576	.43	-.05842	.01114
.19	-.03052	.00605	.44	-.05914	.01125
.20	-.03200	.00634	.45	-.05981	.01136
.21	-.03346	.00662	.46	-.06044	.01145
.22	-.03489	.00689	.47	-.06103	.01153
.23	-.03631	.00717	.48	-.06157	.01160
.24	-.03770	.00743	.49	-.06206	.01167
.25	-.03906	.00769	.50	-.06250	.01172

s	$\binom{s+1}{3}$	$\binom{s+2}{5}$	s	$\binom{s+1}{3}$	$\binom{s+2}{5}$
.50	-.06250	.01172	.75	-.05469	.00940
.51	-.06289	.01176	.76	-.05350	.00916
.52	-.06323	.01179	.77	-.05224	.00890
.53	-.06352	.01181	.78	-.05091	.00863
.54	-.06376	.01182	.79	-.04949	.00835
.55	-.06394	.01182	.80	-.04800	.00806
.56	-.06406	.01181	.81	-.04643	.00776
.57	-.06413	.01179	.82	-.04477	.00745
.58	-.06415	.01175	.83	-.04304	.00712
.59	-.06410	.01170	.84	-.04122	.00679
.60	-.06400	.01165	.85	-.03931	.00644
.61	-.06384	.01158	.86	-.03732	.00608
.62	-.06361	.01150	.87	-.03525	.00572
.63	-.06333	.01141	.88	-.03309	.00534
.64	-.06298	.01131	.89	-.03084	.00495
.65	-.06256	.01119	.90	-.02850	.00455
.66	-.06208	.01106	.91	-.02607	.00413
.67	-.06154	.01093	.92	-.02355	.00371
.68	-.06093	.01078	.93	-.02094	.00328
.69	-.06025	.01062	.94	-.01824	.00284
.70	-.05950	.01044	.95	-.01544	.00239
.71	-.05868	.01026	.96	-.01254	.00193
.72	-.05779	.01006	.97	-.00955	.00146
.73	-.05683	.00985	.98	-.00647	.00098
.74	-.05580	.00963	.99	-.00328	.00050
.75	-.05469	.00940	1.00	-.00000	.00000

Table IV

LAGRANGE'S COEFFICIENTS FOR FIVE

EQUALLY SPACED POINTS

s	$L_{-2}(s)$	$-L_{-1}(s)$	$L_0(s)$	$L_1(s)$	$-L_2(s)$	s
.00	.000000	.000000	1.000000	.000000	.000000	.00
.01	.000829	.006600	.999875	.006733	.000837	.01
.02	.001649	.013065	.999500	.013599	.001683	.02
.03	.002460	.019396	.998875	.020595	.002535	.03
.04	.003261	.025590	.998001	.027722	.003395	.04
.05	.004052	.031647	.996877	.034978	.004260	.05
.06	.004833	.037566	.995503	.042362	.005131	.06
.07	.005602	.043347	.993881	.049872	.006008	.07
.08	.006359	.048988	.992010	.057508	.006888	.08
.09	.007104	.054489	.989891	.065267	.007774	.09
.10	.007838	.059850	.987525	.073150	.008663	.10
.11	.008558	.065069	.984912	.081154	.009554	.11
.12	.009265	.070147	.982052	.089277	.010447	.12
.13	.009958	.075081	.978946	.097520	.011343	.13
.14	.010637	.079873	.975596	.105879	.012238	.14
.15	.011302	.084522	.972002	.114353	.013135	.15
.16	.011953	.089027	.968164	.122941	.014031	.16
.17	.012588	.093387	.964084	.131642	.014927	.17
.18	.013208	.097603	.959762	.140453	.015820	.18
.19	.013812	.101674	.955201	.149373	.016712	.19
.20	.014400	.105600	.950400	.158400	.017600	.20
.21	.014972	.109381	.945361	.167532	.018485	.21
.22	.015527	.113016	.940086	.176768	.019365	.22
.23	.016065	.116505	.934575	.186106	.020240	.23
.24	.016586	.119849	.928829	.195543	.021110	.24
.25	.017091	.123047	.922952	.205078	.021973	.25
.26	.017576	.126099	.916642	.214709	.022828	.26
.27	.018044	.129005	.910204	.224434	.023676	.27
.28	.018493	.131766	.903537	.234225	.024515	.28
.29	.018925	.134381	.896643	.244156	.025344	.29
.30	.019338	.136850	.889525	.254150	.026163	.30

s	$L_{-2}(s)$	$-L_{-1}(s)$	$L_0(s)$	$L_1(s)$	$-L_2(s)$	s
.30	.019338	.136850	.889525	.254150	.026163	.30
.31	.019731	.139174	.882184	.264229	.026970	.31
.32	.020106	.141353	.874621	.274391	.927766	.32
.33	.020462	.143387	.866840	.284634	.028549	.33
34	.020798	.145277	.858841	.294955	.029318	.34
.35	.021115	.147022	.850627	.305353	.030073	.35
.36	.021412	.148623	.842199	.315824	.030812	.36
.37	.021689	.150081	.833560	.326368	.031536	.37
.38	.021946	.151397	.824713	.336979	.032242	.38
.39	.022180	.152569	.815686	.347658	.032930	.39
.40	.022400	.153600	.806400	.358400	.033600	.40
.41	.022596	.154489	.796939	.369204	.034250	.41
.42	.022773	.155238	.787279	.380066	.034879	.42
.43	.022928	.155847	.777422	.390984	.035487	.43
.44	.023063	.156316	.767370	.401956	.036073	.44
.45	.023177	.156647	.757127	.412978	.036635	.45
.46	.023271	.156840	.746694	.424048	.037173	.46
.47	.023344	.156896	.736074	.435163	.037686	.47
.48	.023396	.156815	.725271	.446321	.038172	.48
.49	.023427	.156600	.714287	.457517	.038631	.49
.50	.023438	.156250	.703125	.468750	.039063	.50

Table IV applies to the problem where five equally spaced values of x, x_{-2}, x_{-1}, x_0, x_1, x_2 are given and the point x at which interpolation is to be made is nearer to x_0 than to the other points. If $x > x_0$ the points are to be numbered thus

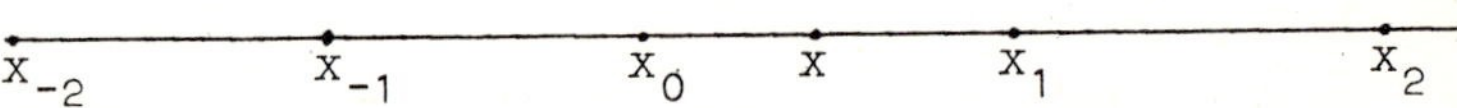

but if $x < x_0$ the points are to be numbered thus

x_2 x_1 x x_0 x_{-1} x_{-2}

In either case $s = \dfrac{x - x_0}{x_1 - x_0}$.

Table V

LEGENDRE'S POLYNOMIALS
(ADAPTED TO THE INTERVAL $0 \leq x \leq 1$)

x	$P_1(x)$	$P_2(x)$	$P_3(x)$	$P_4(x)$	$P_5(x)$	x
.00	1.00-	1.0000	1.00000-	1.00000	1.00000-	1.00*
.01	.98-	.9406	.88298-	.80886	.72045-	.99
.02	.96-	.8824	.77184-	.63489	.47962-	.98
.03	.94-	.8254	.66646-	.47728	.27438-	.97
.04	.92-	.7696	.56672-	.33522	.10175-	.96
.05	.90-	.7150	.47250-	.20794	-.04114	.95
.06	.88-	.6616	.38368-	.09467	-.15699	.94
.07	.86-	.6094	.30014-	-.00534-	-.24838	.93
.08	.84-	.5584	.22176-	-.09281-	-.31774	.92
.09	.82-	.5086	.14842-	-.16847-	-.36739	.91
.10	.80-	.4600	.08000-	-.23300-	-.39952	.90
.11	.78-	.4126	.01638-	-.28709-	-.41618	.89
.12	.76-	.3664	-.04256	-.33140-	-.41931	.88
.13	.74-	.3214	-.09694	-.36659-	-.41074	.87
.14	.72-	.2776	-.14688	-.39327-	-.39217	.86
.15	.70-	.2350	-.19250	-.41206-	-.36520	.85
.16	.68-	.1936	-.23392	-.42356-	-.33131	.84
.17	.66-	.1534	-.27126	-.42836-	-.29188	.83
.18	.64-	.1144	-.30464	-.42700-	-.24819	.82
.19	.62-	.0766	-.33418	-.42004-	-.20142	.81
.20	.60-	.0400	-.36000	-.40800-	-.15264	.80
.21	.58-	.0046	-.38222	-.39140-	-.10285	.79
.22	.56-	-.0296-	-.40096	-.37074-	-.05294	.78
.23	.54-	-.0626-	-.41634	-.34649-	-.00372	.77
.24	.52-	-.0944-	-.42848	-.31912-	.04409-	.76
.25	.50-	-.1250-	-.43750	-.28906-	.08984-	.75

*When x is taken from this column, use sign to the right.

x	$P_1(x)$	$P_2(x)$	$P_3(x)$	$P_4(x)$	$P_5(x)$	x
.25	.50-	-.1250-	-.43750	-.28906-	.08984-	.75*
.26	.48-	-.1544-	-.44352	-.25676-	.13298-	.74
.27	.46-	-.1826-	-.44666	-.22261-	.17301-	.73
.28	.44-	-.2096-	-.44704	-.18702-	.20951-	.72
.29	.42-	-.2354-	-.44478	-.15036-	.24215-	.71
.30	.40-	-.2600-	-.44000	-.11300-	.27064-	.70
.31	.38-	-.2834-	-.43282	-.07528-	.29477-	.69
.32	.36-	-.3056-	-.42336	-.03752-	.31438-	.68
.33	.34-	-.3266-	-.41174	-.00004-	.32937-	.67
.34	.32-	-.3464-	-.39808	.03688	.33970	.66
.35	.30-	-.3650-	-.39250	.07294	.34539-	.65
.36	.28-	-.3824-	-.36512	.10789	.34647-	.64
.37	.26-	-.3986-	-.34606	.14149	.34307-	.63
.38	.24-	-.4136-	-.32544	.17352	.33531-	.62
.39	.22-	-.4274-	-.30338	.20375	.32339-	.61
.40	.20-	-.4400-	-.28000	.23200	.30752-	.60
.41	.18-	-.4514-	-.25542	.25809	.28796-	.59
.42	.16-	-.4616-	-.22976	.28187	.26498-	.58
.43	.14-	-.4706-	-.20314	.30318	.23891-	.57
.44	.12-	-.4784-	-.17568	.32191	.21008-	.56
.45	.10-	-.4850-	-.14750	.33794	.17883-	.55
.46	.08-	-.4904-	-.11872	.35118	.14555-	.54
.47	.06-	-.4046-	-.08946	.36156	.11062-	.53
.48	.04-	-.4976-	-.05984	.36901	.07444-	.52
.49	.02-	-.4994-	-.02998	.37350	.03743-	.51
.50	.00	-.5000-	.00000	.37500	.00000	.50

*When x is taken from this column, use sign to the right.

Table VI

ORTHOGONAL POLYNOMIALS
FOR n + 1 EQUALLY SPACED POINTS

n = 5 (6 points)

s	$P_1(s)$	$P_2(s)$	$P_3(s)$	$P_4(s)$	$P_5(s)$
0	5	5	5	1	1
1	3	-1	-7	-3	-5
2	1	-4	-4	2	10
3	-1	-4	4	2	-10
4	-3	-1	7	-3	5
5	-5	5	-5	1	-1
S_m	70	84	180	28	252

n = 6 (7 points)

s	$P_1(s)$	$P_2(s)$	$P_3(s)$	$P_4(s)$	$P_5(s)$
0	3	5	1	3	1
1	2	0	-1	-7	-4
2	1	-3	-1	1	5
3	0	-4	0	6	0
4	-1	-3	1	1	-5
5	-2	0	1	-7	4
6	-3	5	-1	3	-1
S_m	28	84	6	154	84

n = 7 (8 points)

s	$P_1(s)$	$P_2(s)$	$P_3(s)$	$P_4(s)$	$P_5(s)$
0	7	7	7	7	7
1	5	1	-5	-13	-23
2	3	-3	-7	-3	17
3	1	-5	-3	9	15
4	-1	-5	3	9	-15
5	-3	-3	7	-3	-17
6	-5	1	5	-13	23
7	-7	7	-7	7	-7
S_m	168	168	264	616	2184

n = 8 (9 points)

s	$P_1(s)$	$P_2(s)$	$P_3(s)$	$P_4(s)$	$P_5(s)$
0	4	28	14	14	4
1	3	7	-7	-21	-11
2	2	-8	-13	-11	4
3	1	-17	-9	9	9
4	0	-20	0	18	0
5	-1	-17	9	9	-9
6	-2	-8	13	-11	-4
7	-3	7	7	-21	11
8	-4	28	-14	14	-4
S_m	60	2772	990	2002	468

n = 9 (10 points)

s	$P_1(s)$	$P_2(s)$	$P_3(s)$	$P_4(s)$	$P_5(s)$
0	9	6	42	18	6
1	7	2	-14	-22	-14
2	5	-1	-35	-17	1
3	3	-3	-31	3	11
4	1	-4	-12	18	-6
5	-1	-4	12	18	-6
6	-3	-3	31	3	-11
7	-5	-1	35	-17	-1
8	-7	2	14	-22	14
9	-9	6	-42	18	-6
S_m	330	132	8580	2860	780

n = 10 (11 points)

s	$P_1(s)$	$P_2(s)$	$P_3(s)$	$P_4(s)$	$P_5(s)$
0	5	15	30	6	3
1	4	6	-6	-6	-6
2	3	-1	-22	-6	-1
3	2	-6	-23	-1	4
4	1	-9	-14	4	4
5	0	-10	0	6	0
6	-1	-9	14	4	-4
7	-2	-6	23	-1	-4
8	-3	-1	22	-6	1
9	-4	6	6	-6	6
10	-5	15	-30	6	-3
S_m	110	858	4290	286	156

n = 11 (12 Points)

s	$P_1(s)$	$P_2(s)$	$P_3(s)$	$P_4(s)$	$P_5(s)$
0	11	55	33	33	33
1	9	25	-3	-27	-57
2	7	1	-21	-33	-21
3	5	-17	-25	-13	29
4	3	-29	-19	12	44
5	1	-35	-7	28	20
6	-1	-35	7	28	-20
7	-3	-29	19	12	-44
8	-5	-17	25	-13	-29
9	-7	1	21	-33	21
10	-9	25	3	-27	57
11	-11	55	-33	33	-33
S_m	572	12012	5148	8008	15912

n = 12 (13 points)

s	$P_1(s)$	$P_2(s)$	$P_3(s)$	$P_4(s)$	$P_5(s)$
0	6	22	11	99	22
1	5	11	0	-66	-33
2	4	2	-6	-96	-18
3	3	-5	-8	-54	11
4	2	-10	-7	11	26
5	1	-13	-4	64	20
6	0	-14	0	84	0
7	-1	-13	4	64	-20
8	-2	-10	7	11	-26
9	-3	-5	8	-54	-11
10	-4	2	6	-96	18
11	-5	11	0	-66	33
12	-6	22	-11	99	-22
S_m	182	2002	572	68068	6188

n = 13 (14 points)

s	$P_1(s)$	$P_2(s)$	$P_3(s)$	$P_4(s)$	$P_5(s)$
0	13	13	143	143	143
1	11	7	11	-77	-187
2	9	2	-66	-132	-132
3	7	-2	-98	-92	28
4	5	-5	-95	-13	139
5	3	-7	-67	63	145
6	1	-8	-24	108	60
7	-1	-8	24	108	-60
8	-3	-7	67	63	-145
9	-5	-5	95	-13	-139
10	-7	-2	98	-92	-28
11	-9	2	66	-132	132
12	-11	7	-11	-77	187
13	-13	13	-143	143	-143
S_m	910	728	97240	136136	235144

n = 14 (15 points)

s	$P_1(s)$	$P_2(s)$	$P_3(s)$	$P_4(s)$	$P_5(s)$
0	7	91	91	1001	1001
1	6	52	13	-429	-1144
2	5	19	-35	-869	-979
3	4	-8	-58	-704	-44
4	3	-29	-61	-249	751
5	2	-44	-49	251	1000
6	1	-53	-27	621	675
7	0	-56	0	756	0
8	-1	-53	27	621	-675
9	-2	-44	49	251	-1000
10	-3	-29	61	-249	-751
11	-4	-8	58	-704	44
12	-5	19	35	-869	979
13	-6	52	-13	-429	1144
14	-7	91	-91	1001	-1001
S_m	280	37128	39780	6466460	10581480

n = 15 (16 points)

s	$P_1(s)$	$P_2(s)$	$P_3(s)$	$P_4(s)$	$P_5(s)$
0	15	35	455	273	143
1	13	21	91	-91	-143
2	11	9	-143	-221	-143
3	9	-1	-267	-201	-33
4	7	-9	-301	-101	77
5	5	-15	-265	23	131
6	3	-19	-179	129	115
7	1	-21	-63	189	45
8	-1	-21	63	189	-45
9	-3	-19	179	129	-115
10	-5	-15	265	23	-131
11	-7	-9	301	-101	-77
12	-9	-1	267	-201	33
13	-11	9	143	-221	143
14	-13	21	-91	-91	143
15	-15	35	-455	273	-143
S_m	1360	5712	1007760	470288	201552

n = 16 (17 points)

s	$P_1(s)$	$P_2(s)$	$P_3(s)$	$P_4(s)$	$P_5(s)$
0	8	40	28	52	104
1	7	25	7	-13	-91
2	6	12	-7	-39	-104
3	5	1	-15	-39	-39
4	4	-8	-18	-24	36
5	3	-15	-17	-3	83
6	2	-20	-13	17	88
7	1	-23	-7	31	55
8	0	-24	0	36	0
9	-1	-23	7	31	-55
10	-2	-20	13	17	-88
11	-3	-15	17	-3	-83
12	-4	-8	18	-24	-36
13	-5	1	15	-39	39
14	-6	12	7	-39	104
15	-7	25	-7	-13	91
16	-8	40	-28	52	-104
S_m	408	7752	3876	16796	100776

n = 17 (18 points)

s	$P_1(s)$	$P_2(s)$	$P_3(s)$	$P_4(s)$	$P_5(s)$
0	17	68	68	68	884
1	15	44	20	-12	-676
2	13	23	-13	-47	-871
3	11	5	-33	-51	-429
4	9	-10	-42	-36	156
5	7	-22	-42	-12	588
6	5	-31	-35	13	733
7	3	-37	-23	33	583
8	1	-40	-8	44	220
9	-1	-40	8	44	-220
10	-3	-37	23	33	-583
11	-5	-31	35	13	-733
12	-7	-22	42	-12	-588
13	-9	-10	42	-36	-156
14	-11	5	33	-51	429
15	-13	23	13	-47	871
16	-15	44	-20	-12	676
17	-17	68	-68	68	-884
S_m	1938	23256	23256	28424	6953544

n = 18 (19 points)

s	$P_1(s)$	$P_2(s)$	$P_3(s)$	$P_4(s)$	$P_5(s)$
0	9	51	204	612	102
1	8	34	68	-68	-68
2	7	19	-28	-388	-98
3	6	6	-89	-453	-58
4	5	-5	-120	-354	3
5	4	-14	-126	-168	54
6	3	-21	-112	42	79
7	2	-26	-83	227	74
8	1	-29	-44	352	44
9	0	-30	0	396	0
10	-1	-29	44	352	-44
11	-2	-26	83	227	-74
12	-3	-21	112	42	-79
13	-4	-14	126	-168	-54
14	-5	-5	120	-354	-3
15	-6	6	89	-453	58
16	-7	19	28	-388	98
17	-8	34	-68	-68	68
18	-9	51	-204	612	-102
S_m	570	13566	213180	2288132	89148

n = 19 (20 points)

s	$P_1(s)$	$P_2(s)$	$P_3(s)$	$P_4(s)$	$P_5(s)$
0	19	57	969	1938	1938
1	17	39	357	-102	-1122
2	15	23	-85	-1122	-1802
3	13	9	-377	-1402	-1222
4	11	-3	-539	-1187	-187
5	9	-13	-591	-687	771
6	7	-21	-553	-77	1351
7	5	-27	-445	503	1441
8	3	-31	-287	948	1076
9	1	-33	-99	1188	396
10	-1	-33	99	1188	-396
11	-3	-31	287	948	-1076
12	-5	-27	445	503	-1441
13	-7	-21	553	-77	-1351
14	-9	-13	591	-687	-771
15	-11	-3	539	-1187	187
16	-13	9	377	-1402	1222
17	-15	23	85	-1122	1802
18	-17	39	-357	-102	1122
19	-19	57	-969	1938	-1938
S_m	2660	17556	4903140	22881320	31201800

n = 20 (21 points)

s	$P_1(s)$	$P_2(s)$	$P_3(s)$	$P_4(s)$	$P_5(s)$
0	10	190	570	969	3876
1	9	135	228	0	-1938
2	8	82	-24	-510	-3468
3	7	37	-196	-680	-2618
4	6	-2	-298	-615	-788
5	5	-35	-340	-406	1063
6	4	-62	-332	-130	2354
7	3	-83	-284	150	2819
8	2	-98	-206	385	2444
9	1	-107	-108	540	1404
10	0	-110	0	594	0
11	-1	-107	108	540	-1404
12	-2	-98	206	385	-2444
13	-3	-83	284	150	-2819
14	-4	-62	332	-130	-2354
15	-5	-35	340	-406	-1063
16	-6	-2	298	-615	788
17	-7	37	196	-680	2618
18	-8	82	24	-510	3468
19	-9	133	-228	0	1938
20	-10	190	-570	969	-3876
S_m	770	201894	1730520	5720330	121687020

Table VII

INTEGRALS OF BINOMIAL COEFFICIENTS, $\int_0^s \binom{t}{k}\, dt$

s \ k	0	1	2	3	4	5	6	7	8	9
-1	-1	$\frac{1}{2}$	$\frac{-5}{12}$	$\frac{3}{8}$	$\frac{-251}{720}$	$\frac{95}{288}$	$\frac{-19087}{60480}$	$\frac{5257}{17280}$	$\frac{-1070017}{3628800}$	$\frac{25713}{89600}$
1	1	$\frac{1}{2}$	$\frac{-1}{12}$	$\frac{1}{24}$	$\frac{-19}{720}$	$\frac{3}{160}$	$\frac{-863}{60480}$	$\frac{275}{24192}$	$\frac{-33953}{3628800}$	$\frac{8183}{1036800}$
2	2	2	$\frac{1}{3}$	0	$\frac{-1}{90}$	$\frac{1}{90}$	$\frac{-37}{3780}$	$\frac{8}{945}$	$\frac{-119}{16200}$	$\frac{9}{1400}$
3	3	$\frac{9}{2}$	$\frac{9}{4}$	$\frac{3}{8}$	$\frac{-3}{80}$	$\frac{3}{160}$	$\frac{-29}{2240}$	$\frac{9}{896}$	$\frac{-369}{44800}$	$\frac{25}{3548}$
4	4	8	$\frac{20}{3}$	$\frac{8}{3}$	$\frac{14}{45}$	0	$\frac{-8}{945}$	$\frac{8}{945}$	$\frac{-107}{14175}$	$\frac{94}{14175}$
5	5	$\frac{25}{2}$	$\frac{175}{12}$	$\frac{75}{8}$	$\frac{425}{144}$	$\frac{95}{288}$	$\frac{-275}{12096}$	$\frac{275}{24192}$	$\frac{-175}{20736}$	$\frac{25}{3584}$
6	6	18	27	24	$\frac{123}{10}$	$\frac{33}{10}$	$\frac{41}{140}$	0	$\frac{-9}{1400}$	$\frac{9}{1400}$
7	7	$\frac{49}{2}$	$\frac{539}{12}$	$\frac{1225}{24}$	$\frac{26117}{720}$	$\frac{2499}{160}$	$\frac{30919}{8640}$	$\frac{5257}{17280}$	$\frac{-8183}{518400}$	$\frac{8183}{1036800}$
8	8	32	$\frac{208}{3}$	96	$\frac{3928}{45}$	$\frac{2336}{45}$	$\frac{18128}{945}$	$\frac{736}{189}$	$\frac{3956}{14175}$	0

Table VIII

GAMMA AND DIGAMMA FUNCTIONS

x	$\Gamma(x+1) = x!$	$\psi(x)$	x	$\Gamma(x+1) = x!$	$\psi(x)$
0	1.00000	-.57722	.50	.88623	.03649
.02	.98884	-.54480	.52	.88704	.05502
.04	.97844	-.51327	.54	.88818	.07323
.06	.96874	-.48263	.56	.88964	.09114
.08	.95973	-.45280	.58	.89142	.10874
.10	.95135	-.42375	.60	.89352	.12605
.12	.94359	-.39546	.62	.89592	.14308
.14	.93642	-.36787	.64	.89864	.15983
.16	.92980	-.34095	.66	.90167	.17633
.18	.92373	-.31469	.68	.90500	.19256
.20	.91817	-.28904	.70	.90864	.20855
.22	.91311	-.26398	.72	.91258	.22429
.24	.90852	-.23949	.74	.91683	.23980
.26	.90440	-.21555	.76	.92137	.25509
.28	.90072	-.19212	.78	.92623	.27015
.30	.89747	-.16919	.80	.93138	.28499
.32	.89464	-.14674	.82	.93685	.29963
.34	.89222	-.12475	.84	.94261	.31406
.36	.89018	-.10321	.86	.94869	.32829
.38	.88854	-.08209	.88	.95507	.34233
.40	.88726	-.06138	.90	.96177	.35618
.42	.88636	-.04107	.92	.96877	.36985
.44	.88581	-.02114	.94	.97610	.38334
.46	.88560	-.00158	.96	.98374	.39666
.48	.88575	+.01763	.98	.99171	.40980
.50	.88623	.03649	1.00	1.00000	.42278

For values beyond the range of the tables use the formulas

$$\Gamma(x + n + 1) = (x + n)^{(n)} \quad \Gamma(x + 1)$$

$$\psi(x + n) = \frac{1}{x + n} + \frac{1}{x + n - 1} + \ . \ . \ . + \frac{1}{x + 1} + \psi(x)$$

INDEX